# a guide to self help solar water heating

# PRACTICAL SOLAR H

## a d-i-y guide
## to
## solar water heating

## Kevin McCartney
## with
## Brian Ford

## Prism Press

Published 1978 by

PRISM PRESS
Stable Court,
Chalmington,
Dorchester,
Dorset DT2 0HB

Copyright 1978 Kevin McCartney


Other copyrights are held by the individuals and organisations
listed under 'acknowledgements'.


ISBN 0 904727 53 X Hardback
ISBN 0 904727 54 8 Paperback

Printed by

Unwin Brothers Limited,
The Gresham Press, Old Woking, Surrey.

for

**Marijke and the Sun**

"You give all your brightness away
But it only makes you brighter"

*from Wee Tam by
the Incredible String Band*

## ACKNOWLEDGEMENTS

Brian Ford: Text illustrations.
Annalouise: Cover illustration.
John R. G. Corbett: for permission to reproduce his article on a DIY electronic differential temperature controller.
D.N.W. Chinnery and the Council for Scientific and Industrial Research (South Africa): table in appendix 4.
Country College: Swimming pool photographs in chapter 15.
ICI: Photograph in chapter 15.
Industrial (Anti-Corrosion) Services: Table in chapter 5.
Marley Tile Co.: Information on roof construction.
Met. Office and UK-ISES: Data used in appendix 2.
Pilkington Bros.: Illustrations and tables in appendix 3.
Robinsons Developments: Illustrations in chapter 14.
UK-ISES: Chapter 14 owes much to the contributors to the ISES Conference "Solar Energy for Heating Swimming Pools", January 1977. J. C. McVeigh, P. Brunt, B. McNelis, B. Harocopos, J. F. Missenden and N. A. C. Spelman.

# Contents

## 15 Examples of Completed Solar Systems

Evening class installation — council house retro-fit — solar air heater — farm installation — trickle type solar roof — solar system for a festival — school swimming pool solar installation.

## 16 Buying Solar Collectors

Prices — design — materials — casing — glazing — insulation — unusual designs — solar cowboys — table showing characteristics of a selection of commercial panels — list of manufacturers.

## Appendices

# AUTHOR'S NOTE

This book contains the answers to hundreds of questions I have asked myself during the past few years. Many people have helped to answer them and I am grateful to them all. Some deserve special mention. In particular, Brian Ford, who has not only illustrated the book but also acted as a sounding board for many of my ideas and filled in gaps in my own knowledge. Looking back, much of my early work was made possible by Gerry Foley and students of the Rational Technology Unit at the Architectural Association School. Subsequently, the spirit of cooperation has continued with my colleagues in the Rational Technology Group: Pietro Stellon, Colin and Derek Taylor, Steve Harrison, Graham Brown and Dave Hodgeson.

Practical experiments have been financed by the Richardson Science Scholarship, the J. E. Lessor Award and the Science Research Council. Dermot Mcguigan encouraged me to get the book written and I had advice on specific topics from: Peter Hannon/nuclear power; Vicky Chick/economics; Peter Glass/physics; AWK MacGregor/flow patterns inside collectors. Their advice has been sound and any mistakes which may have crept into my rendering is of course my own responsibility. I have also had help from many people in presenting examples of completed solar systems: Joy and Peter Pratt, Don Gray, S. Pallis, Nick Moore, Robin Clarke, Tony Wiggens and Harold Bland. On the commercial and industrial side of things, I have been kept informed by John O'Connell, Alan Russell-Cowan and Nigel Spelman. Many who took part in classes I taught or organised at the Architectural Association, Havering Technical College and Kingston Adult Education Department have also helped me. Education is indeed a two-way process.

I have also had the benefit of meeting a patient and enthusiastic publisher in Colin Spooner. Finally I am glad to have the opportunity to express my gratitude to my wife's family, the Gerritsma's, and Ada Benting for providing peace, quiet and hospitality whilst I was working on the book in summer 1977, and to my own household and the Vegan Cafe for sustenance and a peaceful environment through 1976 and 1977 whilst I have devoted so much of my personal attention to this book.

K.McC. January 1978

# Foreword

Domestic use of solar energy is no longer a science fiction proposition. The principles behind solar water heating are as simple as those behind drying clothes in the sun.

That is not to say that constructing and installing a water heating system is quite so simple as hanging clothes on the washing line. Still it is within the scope of the home handyman or woman. And, once installed, it will operate automatically for many years, heating water whenever weather permits.

Even in northern latitudes almost half of the annual domestic hot water requirements can be supplied by the sun without any changes in lifestyle. The exact savings will vary, not only with the weather, but also with the size of the collector, the type of system and the amount of hot water required and the time it is used. Energy savings would be increased if hot water consumption was controlled to coincide with the availability of solar heated water — choosing a sunny day to have a bath for example, and giving the solar tank time to heat up by waiting till mid-day before starting to do the laundry.

Experiments in solar water heating were carried out as long ago as 1945 in London. Only cost has held up its application to ordinary homes. Now, the greatly increased price for conventional energy sources, reflecting the rapid depletion of fossil fuel resources, has made solar heating economically competitive in many situations.

Since the energy crisis in 1973, over seventy U.K. companies have begun to market solar collectors. Furthermore, there have been considerable advances in plumbing techniques over the past decades. These have made DIY installation feasible thereby enabling the initial capital cost to be cut by about one third.

For those who would like to do something practical about the energy problem but do not know a thermosyphoning system from a differential temperature controller, this book provides an introduction to the theoretical aspects, descriptions of the hardware involved and a DIY guide to the construction and installation of solar water heating system. For those looking for an off-the-shelf solution, who have been confused by the wide variety in both price and products there is a survey of commercially available equipment. Those who are still undecided might be interested in the sections on economics and how solar collectors can be visually integrated on an existing house. To inspire and encourage those already planning their system, there is a chapter of case histories on systems which have already been installed.

# 1

# A Safe Energy Source

**The Sun God**

Many of our ancestors took a religious view of the sun as the great benefactor and source of all life. This was a soundly based philosophy and formed a solid foundation for a society supported by an agricultural economy in which knowledge of, and reverence for the sun god's nature and movements were closely linked to survival. For then, as now, it was the sun's energy which supported all life — collected by the leaves of green plants, then absorbed as food by the animals.

Mythology however shows that even in ancient times there were men who hoped to enslave the sun for their own ends. Polynesian Indians tell of a hero named Maui who at dawn laid in wait at the Eastern edge of the world, armed with nets and strong ropes to ensnare the rising sun. He is said then to have beaten the sun half to death and as a result the days since have been much cooler.

People of more northerly latitudes are unlikely to thank Maui for the cooler days we have had, and would do well to adopt a more gentle approach to catching the sun. Nowadays piping and sheets of glass have profitably been subsituted for ropes and nets. It would be unfortunate though if the Maui-type attitude were retained and only the implements changed. It is after all, this predisposition for indiscriminate exploitation which has brought us to the environmental and cultural crises which face the industrialised world. Energy shortages are only a symptom of this larger problem.

Bearing in mind the warnings ecologists have been giving us throughout the past decade we would perhaps improve our present-day survival chances if we adopted some of the reverence which the followers of Ra, Aten, Phoebus Apollo and other solar deities displayed. Watching the skies, acknowledging our environment, growing increasingly aware of the fantastic swirl of solar energy which sustains the life on our planet, we can begin to solve the problems which confront us.

## Atoms and Energy

The attempts of the ancients to explain the wonder of solar energy are no more amazing than the current explanations put forward by scientists.

Every thing we see, feel and touch is made up of tiny particles which we call atoms. Physics teaches us that the atoms themselves are made up of still smaller parts — a nucleus of protons and neutrons surrounded by orbiting electrons. It was discovered that even these can be broken down into still smaller components.

Modern physics paints a weird landscape in which apparently solid matter can be seen to consist largely of voids, populated by miniscule particles which behave at one time like pieces of matter, and then at another time like waves, or bundles of energy.

All matter can be seen to be ultimately a concentration of energy, exquisitely ordered in a limited number of patterns to produce the elements which chemistry uses as building blocks to produce all the familiar compounds we find in our daily lives.

The concentration of energy in the nucleus of an atom is intense. Hence its potential for releasing energy is immense. This has been demonstrated in the atom splitting experiments where atoms are forced to collide with other particles at great speed. This can cause the atom to disintegrate. Most of the particles recombine very quickly to form new atoms but a few sometimes escape from their material form thereby releasing vast quantities of energy.

## Nuclear Power Station in the Sky

The same process occurs in the sun. Under conditions of extremely high pressure and temperature, light elements such as hydrogen fuse together to form heavier elements. Such transformations release the torrent of radiation which streams forth in all directions into space.

The flow of energy from the sun is so great that even our tiny planet, safely removed, ninety three million miles from the inferno, intercepts some four thousand million, million kilowatt hours each day. If this were surging through your electricity meter, you would get a bill for £17,000,000,000,000,000 every week! In ten days, we receive more energy from the sun than is contained in all the world's fossil fuel reserves.

## The peaceful atom

The atomic bombs dropped on Hiroshima and Nagasaki in 1943 demonstrated publicly that mankind had learned to release the energy in the atom. Following the awesome success of the war technologists, in demonstrating their ability to kill and destroy at a rate previously undreamed of, the technically advanced nations began to devote the bulk of their energy research funds into nuclear power.

2

In the 1950's, promises of cheap, unlimited supplies of nuclear power within two decades made nuclear energy the cornerstone of all the developed world's future energy policies and the "peaceful atom" was heralded as the basis for a new era of prosperity.

Twenty years later, the energy crisis of 1973 and the spiralling rise in costs have served to underline the failure of the earlier promises. Technical difficulties have meant that the fast breeder reactor has still not even been demonstrated on a commercial scale. The fast breeder through its ability to use uranium fuel more efficiently, is the only type of nuclear reactor which might be considered to have any long term future.

## The nuclear threat

Meanwhile the thermal fission reactors which have been commissioned continue to produce radioactive waste with no satisfactory solution yet devised for safeguarding these materials some of which will remain a health hazard for twenty thousand years. Even without major accidents occurring, so-called acceptable leakages from nuclear power plants will raise the background radiation level to which the whole population is subjected. American doctors Tamplin and Gofman have estimated that this could lead to an increase of 32 thousand in the annual U.S. cancer death toll.

Apart from the threat to health, proliferation of atomic fuels such as plutonium will form a threat to peace as they constitute the raw material for the fabrication of atomic weapons. The danger of terrorist abduction during the transportation of such materials will increase the necessity for security operations which are likely to further infringe upon individual liberties. Britain for instance, already has a force of over four hundred special constables, armed with submachine guns, who are accountable, not to parliament, but to the Atomic Energy Authority. Their behaviour is considered to be part of the day to day running of a nationalised industry and therefore not subject to M.P.'s questions.

## The case for a solar weapon

It is perhaps interesting to surmise that, had the Americans been able in 1945 to orbit a giant solar collector which could beam down microwaves capable of razing Hiroshima to the ground, we might now be living in a much less polluted environment, members of a society supported by the unending flow of energy which emanates from the centre of our planetary system.

As things are without the massive state investment in solar technology that a solar weapon would have encouraged, we must look to individuals, and groups of individuals, to realise the importance of establishing an energetic basis for our society which is sustainable long into the future and does not constitute a threat to the very population it is designed to support. It is important that we apply alternative energy sources in our personal environments and attempt to influence public decision making bodies in any policies affecting energy usage.

3

# 2

# Utilisation of Solar Energy

Deliberate indirect utilisation of solar energy began with agriculture. Our farming forefathers selected, protected and nourished certain plant species which they considered useful. These plants converted a small proportion of the solar energy falling on their leaves into chemical energy which held together the compounds which served as food or clothing for humans. This conversion of energy in the plant is called photosynthesis, and is carried out by chlorophyll—the substance which gives plants their greenness. Sunlight enables chlorophyll to fix carbon dioxide from the air, and water from the soil to produce carbohydrates. These carbohydrates constitute a biological energy source for the plant, or for an animal which might eat the plant. All life depends on photosynthesis. All life energy has been transmitted from the sun via plants.

## Five Star Solar Energy

When our ancestors learned to handle fire, there began a new line of development in the indirect utilisation of solar energy. This has culminated in rockets carrying men to the moon, centrally heated houses and all manner of 'horseless carriages'. Since setting light to a few twigs and releasing the solar energy which a tree had collected and stored over a period of years, there has been a fascination with the potential for discharging large quantities of concentrated energy.

The discovery of the fossil fuels, coal and petroleum, has fundamentally influenced historical development.

With a flick of a switch, modern man turns on a fire and spills out solar energy which shone down on the lush vegetation of the Carboniferous forests several hundred million years ago. It became buried with the dead plants which were eventually transformed under great pressure from vegetable waste into the mineral deposits which we mine today.

Our present day society has become addicted to this energy concentrate—a candybar diet of instant energy. Exploitation of the energy deposits has been so rapid however that after only a couple of centuries of feasting, the stockpile which took millions of years to accumulate, will soon be depleted. Long before the fuel reserves are actually exhausted of course, price increases will make them inaccessible to the majority of the population.

It is therefore a clear necessity that we begin to utilise solar energy more directly. To make use of the energy as it arrives rather than relying totally on our prehistoric vintage hoard.

## Energy Crops

One approach is to grow plants specifically for the purpose of producing energy, either by burning the dried vegetable waste, or by distillation to produce a convenient liquid fuel. In Brazil for example, there are plans to cultivate vast areas at present covered by tropical rain forest, and distill the crops to produce ethanol which will be mixed with petroleum for use in conventional internal combustion engines. It is hoped, in this way, to halve the nations oil imports by the turn of the century. The difficulty with such energy crops is the vast areas of land they require. When compared with machines, a plant is not very efficient if looked upon as an energy convertor. Typical efficiencies for the photosynthetic process in plants are less than 1%. Only a small proportion therefore, of the energy which falls on the plant, is actually captured. To supply the energy requirements of a city like London, would require an area equivalent to two-thirds of the agricultural land in England and Wales. The cost of harvesting and transporting energy crops must also be taken into account.

## Solar Cells

Scientists working on space research have developed a much more direct means of converting the sun's energy into a convenient form. When they expose a sandwich of two thin layers of silicon crystals to the sun, an electrical potential difference is created between them. This is achieved by mixing minute quantities of another two elements in the layers. For example, arsenic can be added to the upper layer. This provides a 'loose' electron which is liberated by incoming radiation. Boron mixed in the lower layer would tend to attract the electron and hence the voltage which can be measured between the two layers. If enough of these cells, which are usually circular and measure 50—75 mm. (2—3 inches) in diameter, are connected together, a useful electrical current can be obtained.

Solar cells, or photovoltaic cells as they are sometimes called, are however very expensive. Converting solar energy into electricity at an efficiency of about 10%, the whole roof would have to be covered if they were to supply a domestic load. Such an installation would cost tens of thousands of pounds even before providing for electrical storage batteries. Their applications at present are confined to situations where there is no mains supply and the delivery of fuels is difficult. They are used in satellites, manned space crafts and remote navigational beacons.

## Concentrators

If electricity is to be produced from solar energy at competitive prices in the future, it will probably be achieved with the aid of concentrators. These are quite simply, highly reflective surfaces,

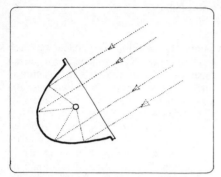

2.1 Parabolic Concentrating Collector —
cross-sectional view showing focussing
of the sun's rays.

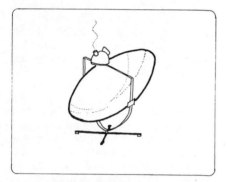

2.2 Parabolic Reflector used for cooking.

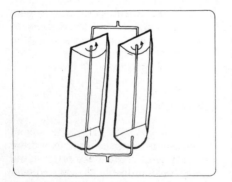

2.3 Cylindrical Concentrator with mobile
reflectors and stationary absorber pipes.

shaped and orientated so as to focus the
radiation falling on a large area, onto a
much smaller area where it will be converted into electricity by solar cells, or
absorbed as heat to produce steam,
which will in turn drive a generator.
This is a likely development because the
reflectors are cheaper than the larger
area of cells or absorbers which they
replace. Furthermore, in the case of
steam generating plants, concentration
of the sun's energy is necessary in order
to achieve the required temperature in
the absorber. Higher temperatures are
possible in concentrators because the
absorber area, the only part which gets
hot, is small in comparison to the area
from which energy is being collected. As
heat loss is proportional to surface area,
the relatively smaller absorber will
suffer less heat loss.

The huge concentrator system covering
a hillside at Odeillo in the French Pyre-
nees, reaches temperatures of over 4000
deg. C. and is used to smelt metals.
More modestly proportioned concentrators have been built to serve as cookers
in the third world. The ideal form for a
concentrator is a shallow dish shaped
parabola. In a cooker, the absorber can
be simply a black pot. Where a large
quantity of liquid must be heated to
high temperatures, it can be circulated
through a pipe which runs along the
focus of a trough shaped concentrator.

The difficulty with all these concentrating devices is that they need to follow
the sun as it moves across the sky. A
small cooker can be moved by hand, but
a larger power producing device requires
an automatic tracking engine. An additional problem is that it is not possible
to concentrate the scattered radiation
which comes through clouds. In many
countries, a large proportion of the
annual total of solar energy does come
through the clouds; in Britain, for
example, more than half.

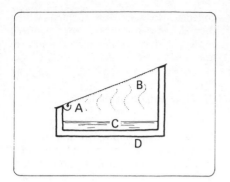

2.4 Solar Still
A Distilled water collected in gutter.
B Condensation gathering on underside of glass cover.
C Impure water.
D Black container.

2.5 Trompe Solar Wall — black painted south wall covered with glass.
A Winter heating mode.
B Summer cooling mode.

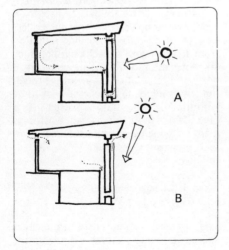

A concentrating system will operate only in direct sunshine. In countries with less than 1500 hours of bright sunshine they will be redundant so much of the time that they will be uneconomical.

**Solar Stills**

Another use to which solar energy is put in sunnier climates is the distillation of water. Brackish water is passed through blackened troughs under sloping glass covers, where it is allowed to evaporate in the sun. Salt and other impurities are left in the troughs whilst the distilled water is condensed on the cold underside of the glass. From here it runs off into channels along the outer edge of the troughs which lead it to a storage tank.

**Solar Cooling**

To use the sun for cooling buildings must be one of the most resourceful applications of solar energy. The simplest method used is to create a strong convection current in the air which will draw a draught of fresh air through the building. This has been done by covering the south facade of the building with glass and leaving an opening at the top. Air between the glass and the wall will rise when heated and will escape through the opening. If ducts are made through the base of the wall, the rising current of air will suck air from the interior through them and hence draw fresh air into the building from windows in the shaded north side.

More complex solar air conditioning systems have been stalled in experimental houses. These are not yet economically competitive with conventional refrigeration installations. They operate either by using high temperature solar

7

collectors to power a heat engine which then drives a compressor in a standard cooler, or they are based on the absorption refrigeration cycle as in a gas fridge.

## Flat Plate Solar Collectors

The solar installations which will most rapidly pay back the cost of their construction, in the form of reduction in fuel bills, are those which serve a requirement for low grade heat, i.e. where temperatures below 80 degrees C. are sufficient. Such temperatures can be attained in simple flat plate solar absorbers, insulated in weatherproof casings, glazed and mounted to face the sun's midday position. They have no need of complex tracking mechanisms and they can also make use of some of the scattered radiation which is so abundant on bright but cloudy days. Flat plate solar collectors can be used to heat water for washing, swimming pools and some industrial processes. In more complex systems, they can also be used for space heating.

The basic flat plate solar system has three functions: collection, circulation, and storage. There follows a chapter on each. Briefly one can say that the collection takes place by absorbing radiation from the sun with a black coating. Once absorbed the radiant energy converts to heat, the circulation of a fluid through the absorbing plate allows the heat to be withdrawn. If the fluid is passed into a tank, this then serves as a storage medium which can store the heat until such time as it is required.

## Space Heating

The greatest difficulty in using solar energy for space heating arises from the

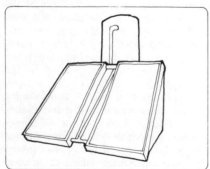

*2.6 Simple Solar Water Heating System — Flat plate collectors with hot water storage tank.*

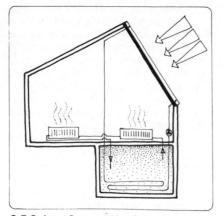

2.7 Solar Space Heating System — Flat plate collectors mounted on roof with large underground storage tank.

obvious fact that the need for heating is greatest when the sun is absent. A solar space heating system therefore has to have an auxilliary heating system to maintain comfortable room temperatures during long periods without sun, or it must have a huge storage tank which will enable heat collected during summer to be held over for use during winter. Such a storage tank would be as large as the house it served and is therefore a very expensive item. Most systems which have been built make a compromise and include a storage capacity of about 3—6 cubic metres (approx. six thousand gallons). When topped up with hot water, this is large enough to supply a house's heating requirements for two or three days. Some form of auxiliary heating is required with such systems.

A more economic means of using solar energy to reduce a building's heating requirements is to exploit the greenhouse effect. Gardeners are familiar with the fact that a glass enclosure can become very warm whenever there is sun, even on cold winter days. Greenhouses can be built on the south side of houses providing an area for food production as well as a source of warm ventilation air.

2.8 Passive Solar Heating — a lean-to greenhouse on the south wall supplies warm air to the building.

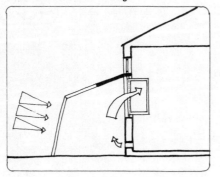

Alternatively, the south wall can be glazed with only a small space between glass and wall, as in the Trompe Wall described in the section on cooling. This same construction can serve to heat the building by closing the opening at the top of the glass and allowing the rising warm air to enter the building through ducts in the top of the wall. If the duct is closed, there will be no air circulation to carry the heat away and heat will be absorbed, therefore, by the wall itself. If it is a thick solid wall, say 300mm (1') of concrete, the heat will take several hours to pass through to the internal surface. Hence the wall would not begin

9

to radiate heat into the room until evening at which time it is most likely to be needed.

The simplest solar collector of all is a south facing window. Depending upon their size, they can contribute some 10—20% to a house's heating requirements provided that heavy curtains or insulating shutters are used to reduce the nightime heat loss through them.

**Solar Water Heating**

Hot water is required throughout the year. Solar water heating systems therefore can operate usefully during the summer when the energy supply is at its greatest. This is why water heating is the most cost effective application for solar systems and why they will be the first to be seen in widespread use. Already, in countries like Israel, one in four apartments are supplied by flat plate solar water heating systems such as those described above. These systems are described in depth in the chapters which follow in the hope that a more widespread familiarity with the simple techniques, and common materials involved will enable more people to benefit further from the sun.

# 3

# Basic Principles

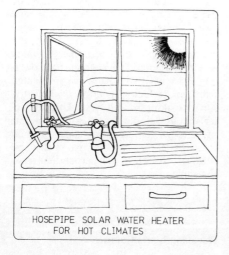

HOSEPIPE SOLAR WATER HEATER
FOR HOT CLIMATES

Collecting solar energy can be very simple. In the sunniest parts of Australia some householders simply snake a black hose pipe in a long loop from their kitchen window, round the yard and back to the sink. And that is all that is necessary to heat their water in the summer.

Where the climate is less favourable, solar collection systems need to be somewhat more sophisticated. They have to put to use as much of the available sun as is practicably possible, collecting it more efficiently and having the capacity to store it for at least the few hours between a sunny morning and the period of peak demand in the evening.

To understand how such improvements can be brought about, it is helpful to consider how energy, particularly heat, moves about and how radiation from the sun interacts with objects which lay in its path.

11

## Heat Transfer

Heat can be transferred between objects in three different ways: conduction, convection and radiation. Conduction takes place when heat travels from one molecule to another through direct physical contact. It is through conduction that the handle of a pan can become hot when it is sitting on the stove. Convection takes place only in fluids. Liquids and gasses expand when heated. This means that each unit volume of the material becomes relatively lighter. The heated fluid therefore tends to rise above surrounding cooler fluids. In this way heat is lifted away from a hot object in what we call convection currents. This can be seen in the clouds of steam which rise above a heated pan of water.

## Radiant Energy

The sun radiates energy in a similar manner to an electric fire. Radiant energy, unlike conducted and convected energy, does not require a physical medium through which to flow. It can pass through a vacuum carrying energy from a hot object to a cooler one. Hence solar energy can cross the gulf of space to bring warmth to the earth. It moves in waves like the ripples which spread out from a pebble dropped in a pond. Very hot objects radiate energy in such a way that there is a very short distance between the crest of each wave, i.e. in short wavelengths. Cooler objects radiate in longer wavelengths. As will be explained later, this difference in the wavelength of radiated energy gives rise to the 'greenhouse effect' which can be of great assistance to systems designed to collect solar energy.

## Colliding with a Sunbeam

When an object lies in the path of radiant energy three things can happen: energy can be transmitted through the object, reflected away, or absorbed by it. The reaction depends upon the physical characteristics of the object, the wavelength of the radiation and the angle of collision.

## Transmission

Transparent materials like glass are good transmitters of solar energy. If solar radiation approaches a sheet of glass at right angles, nearly all the energy will pass through it. Most of the energy which is not transmitted is reflected away. The more oblique the angle between the glass and the sun's rays, the greater the proportion of energy which will be reflected. This explains the dazzling light which is often seen reflected from south-facing windows when the sun is sinking in the west. This characteristic also indicates that the amount of energy entering a glass covered solar collector can be maximised by tilting it so as to be perpendicular to the sun's rays.

## Wavelength and Colour

Solar radiation consists of many different wavelengths. Most of the sun's energy arrives in very short wavelengths which scientists measure in units called microns. There are ten thousand microns in one centimetre. Surprisingly enough our eyes are able to distinguish between wavelengths whose lengths are different only by a fraction of a micron. Few of us though, describe what we see in terms of microns. Rather we use words like violet, indigo and blue for the shortest wavelengths and red, orange and yellow for the longer wavelengths. Not all solar radiation is visible however. The infra-red component has a wavelength too long for our eyes to register, and the ultra-violet is too short.

## Reflection and Absorption

All the colours we see in a landscape are components of solar radiation being reflected in accordance with their wavelength by different materials. The difference between a yellow daffodil and a red rose in terms of their interaction with sunlight is that the daffodil absorbs most of the radiation with the exception of those wavelengths which we perceive as yellow. These are reflected from the petals and into our eyes. The rose on the other hand absorbs the yellow wavelengths but reflects the red. Another rose, with a different chemical structure might reflect nearly all the different wavelengths. When all the colours are reflected together, we see white. In the opposite case, where nearly all the wavelengths are absorbed, we see black objects.

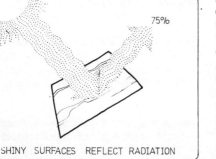

75%

SHINY SURFACES REFLECT RADIATION

Knowing this, we can actually increase the amount of energy an object absorbs simply by giving it a black surface coating.

When radiant energy is absorbed it changes into heat. One might imagine the solar rays, having beamed ninety-three million miles through space, suddenly crashing into an absorbing

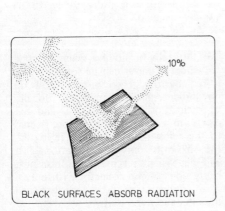

10%

BLACK SURFACES ABSORB RADIATION

13

object. The 'shock' waves set the electrons in the molecules of the absorber vibrating more rapidly. If this shaking up of the electrons is violent enough, we can actually sense their movement as warmth, and measure the change as a rise in temperature.

## Hot and Cold Cars

The effect of different coloured surface finishes on energy absorption can be felt. If you see a white motor car parked alongside a darker one in the sun, lay a hand on the bodywork and the temperature difference should become obvious.

## Re-Radiation

Once the temperature of the absorbing object begins to rise, it also begins to radiate outward. If the absorbing object is intended to function as a collector of solar energy, this re-radiation must be kept to a minimum. This can be done in two ways:
a) exploiting the greenhouse effect.
b) maintaining a low operating temperature.

## The Greenhouse Effect

The greenhouse is like a heat trap. It works because glass, and some plastic materials, act like radiation filters. They transmit a large proportion of solar energy which is mostly short wave radiation, but they do not transmit the

longer waves which are re-radiated from objects heated in the sun. Thus if an absorbing object is covered by a sheet of glass, heat will build up and we will obtain the higher temperatures which gardeners experience in their glasshouses.

## Operating Temperature

If the temperature of the absorbing object is kept low by rapidly extracting heat and transferring it to a store, the amount of energy re-radiated will be correspondingly lower.

When people hear that you are working with solar energy, one of the first questions they will ask you is "What kind of temperatures do you get?"

I have measured temperatures over boiling point in single glazed flat plate solar collectors when the water is not circulating, and I know of others who have measured temperatures as high as 185°C in solar collectors which have been emptied of their water content. When a collector is connected to a storage tank and water circulated through it, its temperature is more likely to be somewhere below 80°C.

All this talk of boiling water in solar collectors however can be something of a red herring. High temperatures do not indicate an efficient system. Ideally of course one would like to end the day with a tank full of water heated to the temperature at which it is used (45-55°C in domestic situations). It will be easier to achieve this however, if we operate the collector with only a small temperature difference between the inlet and the outlet.

A solar collector operating at 30°C is collecting energy more efficiently than a similar collector operating at 60°C. At first this seems ridiculous, but after a little thought it makes sense.

The major source of inefficiency in a solar collector is the heat losses which occur after energy is absorbed. The hotter an object becomes, the greater will be the rate at which it loses heat. A collector operating at the lowest useful temperature will therefore function at its highest efficiency.

The operating temperature can be changed by varying the rate at which water flows through the collectors and by altering the connecting patterns in an array of collectors.

The flow rate can be increased, and hence the operating temperature lowered by reducing the circulation resistance, through using larger diameter piping and minimising the length of the pipe runs and the number of bends. With a variable speed pump, the flow rate can be altered at the turn of a switch.

The effect of different connection patterns between solar collectors is discussed in detail in chapter 10. Briefly it is worth remarking that if the warm outlet from a collector is piped into the inlet of another, the operating temperature of the second will obviously be higher and hence its efficiency will be lower.

## Tracking the Sun

The maximum amount of radiation is intercepted if the absorber is at right angles to the rays. The more oblique the angle, the more energy will be missed — that is to say the radiation actually falling on the object will be less intense.

In an ideal solar collector, we might want the absorber to track the sun on its changing path across the sky, its tilt varying as the sun rises to, and falls from, its midsummer zenith. The mechanism for such a tracking process would be unjustifiably expensive and complex. When using flat plate absorbers which do not rely upon any focussing of the sun's rays, satisfactory results are obtained with a stationary mounting if it provides a tilt and orientation within the limits suggested below.

## Tilt Angle

The tilt angle is the angle between a collector surface and the horizontal. It has often been written that the ideal tilt angle for a solar collector is equal to the latitude of its location plus 10 or 15 degrees. In cloudy climates this can be misleading. In London for example which is situated at a latitude of 52°N the collector which would receive most

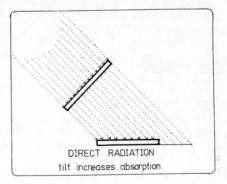

DIRECT RADIATION
tilt increases absorption

15

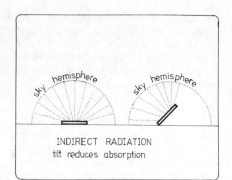

INDIRECT RADIATION
tilt reduces absorption

energy throughout the year would have a tilt of only 34°. This discrepancy is due to the fact that almost half the solar energy received comes not directly from the sun, but is diffused by the cloud cover. This diffuse radiation emanates from the whole sky hemisphere. The ideal angle therefore for a collector aimed at diffuse radiation would be horizontal. This accounts for London's annual optimum tilt of 34° being much flatter than its latitude would lead one to expect. When we are considering an application like space heating the demand of the system will be at a maximum in winter when the sun is low in the sky. In such cases a tilt angle greater than the latitude angle would be appropriate. Similarly, a steep angle would be appropriate in an autonomous house where normal heating is not available.

In such a case it is necessary to optimise the system for the winter months when there is much less sun available. In ordinary housing however, where the intention is to maximise the saving of gas, oil or electricity for water heating, it is best to optimise the system's performance for autumn and spring. In the U.K. this would require a tilt angle of about 45°.

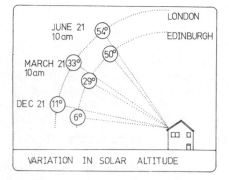

VARIATION IN SOLAR ALTITUDE

This discussion of optimum angles may be irrelevant if you are hoping to install solar collectors on an existing pitched roof. In this case, altering the pitch or constructing special supports is unlikely to be worthwhile. If the existing roof has a pitch of between 30 and 60 degrees, then it is altogether suitable. With steeper or shallower roof slopes, there will be some reduction in the amount of energy collected, this could be compensated by utilising a larger area of collector surface.

## Orientation Angle

Although it is obviously desirable to have the solar absorber facing south, it can in fact be orientated as much as 30 degrees east or west of south with only a negligible reduction in the amount of energy received. Beyond 40 degrees away from south the reduction becomes more marked, particularly in the winter months.

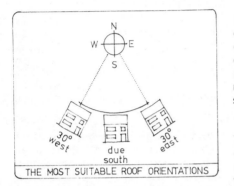

THE MOST SUITABLE ROOF ORIENTATIONS

For absorbers with steep angles of tilt, the losses encountered when facing away from south will be slightly greater. Facing away from the sun's midday location, they rely much more on diffuse radiation. Being so steeply tilted however they turn their back on as much as half the radiating sky hemisphere. A southerly orientation is therefore more critical with vertically mounted absorbers than those mounted on pitched roofs.

## Surface Area

Solar energy can be considered as an infinitely large waterfall pouring itself across the earth's surface. If we are attempting to intercept this flow, it is obvious that the larger the area of the interceptor, the greater will be the catchment. So it is with solar absorbers.

There is though a limit to the amount of surface area which will be useful in any particular situation. This is because a large area which would prove useful in winter, would be largely redundant in summer, producing more heat than could be used, perhaps even requiring cooling.

For domestic water heating an easy rule of thumb has been developed: provide one square foot of absorber surface for every gallon of water to be heated. In practice this might be increased to as much as 1.3 gallons per square foot (one square metre for every 50 to 63 litres).

The average family uses about 160 litres of hot water each day or 60 litres per person per day. Individuals, of course, vary greatly in their use of hot water. Collectors on family houses are usually 3 to 6 square metres in area, smaller ones usually being considered uneconomic on account of the ancilliary equipment, such as pumps and piping, the cost of which is virtually the same for any size of collector.

## Surface Area & Storage

Whilst a large surface area is beneficial for the collection of energy, it is a positive disadvantage for the storage of energy. Heat is lost in proportion to the surface area of the storage vessel. A flat plate of large surface area which is an ideal form for an absorber is the worst form for storage.

A similar contrariness between the functions of collection and storage arises with respect to the characteristic of thermal mass. To optimise both functions it is necessary to separate them.

## Thermal Mass & Reaction Time

Thermal mass is a term used to describe an object's capacity for storing heat. Matter with a high thermal mass may therefore be useful for storing large quantities of heat in a small volume. It will however require a comparatively long period of exposure to sunshine in order to experience any considerable rise in temperature. Matter with a high thermal mass is typically dense, like water and granite. If as suggested the functions of collection and storage are separated low thermal mass in the absorber can be advantageous, as this would allow a more rapid temperature rise. In a solar collector system, it is the temperature rise in the absorber which begins the circulation process which transports absorbed energy into the storage vessel. A low thermal mass will therefore mean a quick reaction time.

In sunny conditions this property is not so important. Once circulation has begun, it will continue throughout the day. In conditions where periods of intense radiation are followed by dull spells, a collector of high thermal mass might have been absorbing energy for several hours and be just reaching the temperature where circulation would begin when a layer of heavy cloud causes a sudden drop in the level of radiation. If the surrounding air temperature is low the temperature would gradually fall again as heat is lost to the environment.

## Conclusion

This chapter has introduced some of the factors which underlie the operation of a solar system, but which cannot normally be seen.

The next three chapters describe the hardware required and detail the function of the three major components of solar water heating installation: the collectors, the circulation system and the heat store.

# 4

# Solar Collectors

Solar collectors are to a solar water heating system what a boiler is to a conventional system. It is in the collectors that the water is heated. The water is circulated through the collector so that solar heat, trapped by an absorber plate, might be transferred to it by physical contact.

The collector consists of four elements:
 i) absorption plate
 ii) translucent cover
 iii) insulation
 iv) casing

The case is in the form of an open-top, shallow box, insulated on the bottom and sides. Within lies the absorber, and above it, like a window, is the translucent cover. If the collector is integrated in the roof of a building, the roof rafters may take the place of the casing. In applications where the water does not need to be heated more than 6°C above the air temperature and the collector site is not exposed to strong winds, an absorber plate alone, uninsulated and uncovered might prove adequate.

## HOW IT WORKS

When solar radiation reaches the cover, three things happen. A small quantity of the energy is absorbed by the cover itself; some is reflected away; the remainder is transmitted through to the absorber plate below. Most of this transmitted energy will be absorbed and transformed into heat which is conducted through the plate to the water channels which are either integral or bonded to it. If water is then circulated through the absorber plate, the collected energy can be extracted.

Unfortunately not all of the absorbed energy can be extracted. For as soon as the absorber plate's temperature begins to rise, it will also begin to lose heat to its surroundings. Heat will be conducted through the casing with which it is in contact. The air in contact with the plate will become heated and rise in a natural convective current thus carrying away more heat. Finally the plate will radiate energy outward at a rate which will increase with its temperature.

## EFFICIENCY

When people talk about the efficiency of a collector they are referring to the quantity of energy extracted by the circulating stream of water expressed as a fraction of the total amount of solar energy falling on the cover of the collector. Hence, on a day when 4 kWh. of solar energy was received on the cover, and the collector contributed 2 kWh. towards heating water, the collector can be said to have operated at 50% efficiency. The efficiency of a collector is not a constant value. It changes with air temperature, wind velocity, solar intensity and the water temperature in the collector. Domestic solar water heaters usually operate at efficiencies between 30% and 50%.

From the description of how the collector works, it can be seen that several things might be done to ensure that the efficiency is as high as possible. The energy gains must be maximised and the heat losses minimised.

In other words we should maximise:
a) transmission through the cover
b) absorption by the plate surface
c) heat transfer from the absorber to the water

At the same time we should minimise:
d) heat conduction through the casing
e) Convection currents on the surface of the absorber
f) re-radiation losses from the plate

## TRANSMISSION THROUGH THE COVER

The proportion of solar energy which passes through the cover depends upon the physical properties of the material used and the cleanliness of its surface.

### Glass

A clean sheet of ordinary window glass transmits about 84% of the energy it receives. Sometimes a special glass with a low iron content is selected because this will absorb less radiation. The actual increase in transmitted energy however is not large enough to justify the increased cost of the glass. It may be worth noting however that a high iron content, and hence a lower transmissivity, is characterised by a greenish tint in the glass.

### Plastic Covers

Many plastics, such as clear acrylic, PVC and GRP, are capable of transmitting as much, or even more energy than glass. Plastics however are on the whole prone to deterioration in their transmissivity when exposed to the ultra violet light of the sun. Also, many of them fail to achieve the greenhouse effect explained in the previous chapter. This is because they transmit not only solar energy, but also the long wave energy which is re-radiated from the absorber when it becomes heated. Finally plastics offer no cost savings when compared with glass unless they are used in thin films. The use of thin films can lead to difficulties in mounting. There have been problems in the past of plastic films flapping in the wind and hence increasing convection heat losses. The manufacturers of Tedlar, one of the most appropriate types of plastic for use in solar collectors, recommend mounting their material stretched across a frame and then permanently heat-shrinking it in an oven. Tedlar film should have a life of 5 - 7 years although there may be problems of the cover tearing if handled roughly during mounting. Its cost is about half that of window glass. A new combination of Tedlar film and rigid polyester sheet may become popular in the future. It is unlikely to be cheaper than glass, but it offers manufacturers the possibility of transporting covered collectors without fears of breakages.

## Double Glazing

Two layers of glass are sometimes used in order to reduce the heat losses from the absorber. This however will also reduce the amount of solar energy entering the collector. A double layer of 4 mm glass will transmit only 71% of the radiation it receives. Double glazing is only beneficial therefore in those circumstances where there are very high heat losses, i.e. where water is being heated to more than about 35°C above the temperature of the outside air. Where double glazing is used, special attention must be paid to the fixing of the inner layer. This will experience much higher temperatures than glass in a normal window and it must have space to expand when heated. Otherwise it will crack.

## Dirt

Dust and grease gathering on the cover will reduce the amount of energy transmitted. This problem will be minimised by the use of glass due to its hard smooth surface. Furthermore a steep mounting angle will assist in the cleansing effect of rainfall. Even with a tilted sheet of glass however the transmissivity is likely to drop by about 20% if it is not cleaned regularly. The exact amount will depend on the amount of dust and pollutants in the local atmosphere.

## ABSORPTION BY THE PLATE SURFACE

The absorption of radiation by the plate is increased by coating it with a matt black surface. So treated it will absorb 80-98% of the radiation reaching it. The surface finish should be as thin as possible so as not to form a barrier to the heat flowing from the exterior to the interior where it will be transferred to the circulating stream of water.

## HEAT TRANSFER FROM ABSORBER TO WATER

Heat transfer at this point can be assisted by constructing the absorber plate from materials which are good thermal conductors and by maximising the contact surface area between the water and the water channel, and between the water channel and the absorber plate.

It is principally in the provision for conducting heat from the absorber into the circulating water that absorber plate designs differ. There are three main types:
  i) Trickle or open trough
 ii) Sandwich or radiator
iii) Tube-and-sheet

## Trickle Absorber

This is the simplest and cheapest type of solar absorber. It consists of a sheet of corrugated roof decking with a perforated water feed pipe running along its upper edge, and a gutter pipe along its lower edge. The holes in the upper pipe, the sparge pipe, are situated so that water pours into the gulleys formed by the concave corrugations and then trickles down them to be collected again by the gutter pipe at the bottom. In this way water can remove heat which the metal panel has absorbed during exposure to the sun.

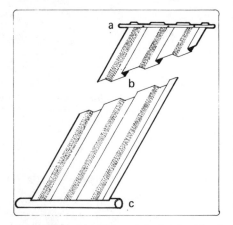

*4.1 Trickle Absorber: a) perforated feed pipe; b) corrugated sheet; c) gutter pipe.*

Such an absorber should be able to efficiently conduct heat from the convex corrugations or ridges, to the concave corrugations where the water flows. To achieve this, the corrugations should be closely spaced and good thermal conductors. Of the commonly available corrugated materials, aluminium is the best choice. Its thermal conductivity is four times greater than that of mild steel. This means that a steel sheet would have to have its corrugations spaced at a quarter the distance of that on an aluminium panel, or it would have to be four times as thick, in order to match the performance of the aluminium in heat transfer.

The surface contact area can be increased by adding detergent to the water. This will decrease the water surface tension which tends to hold the water stream together, preventing it from spreading laterally across the corrugated channels.

A long lasting and robust coating is essential for this type of absorber due to the harsh environment which it must withstand: the corrosive effect of air and water combined and dry temperatures which may rise to almost 200°C causing expansion and rapid contraction when water begins to circulate. A stoved-on finish is probably the safest choice.

Having said that the trickle absorber is the cheapest, it is worth pointing out that it does have several disadvantages when compared to those with sealed waterways. It is less efficient because heat is lost when water running down the open channels evaporates. This evaporation causes a second problem in that it condenses on the comparatively cold surface of the glass cover, misting it and thereby reducing the amount of solar radiation which can enter the collector. Precautions must be taken to ensure that the condensation is not able to cause any damage to the roof on which the absorber is mounted. Dust and dirt is likely to enter the system so that even with filtration of the circulating water, access for cleaning the sparge pipe must be possible in case of blockages. Finally, there is more difficulty in accurately controlling the pump for this type of absorber due to the large fluctuation between the plate's wet and dry temperatures.

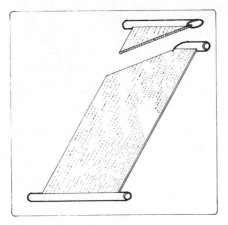

## Sandwich Absorbers

In the sandwich type, water is 'spread' between two sheets of material, the upper of which forms the solar absorbing surface. In this case, water is in contact with the whole, or nearly the whole, of the absorbing surface. Hence heat only has to travel through the thickness of the absorber plate wall to reach the water. With this type then, the thermal conductivity of the materials used is not so important as long as they are not very thick. Hence one finds solar absorbers of the sandwich type being manufactured from a variety of plastics.

When more conductive materials are used they are sometimes bonded together in such a way that water is no longer in direct contact with the whole of the absorber surface. Roll-bond aluminium absorbers and some types of central heating radiators used as absorbers fall into this category which thermally behave partly as sandwich types, and partly as tube-and-sheet types.

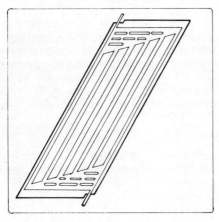

A critical design characteristic in sandwich absorbers is their water capacity. As explained in chapter 3, a high water content will mean a high thermal capacity which in turn results in a slow reacting solar collector. Ideally a solar absorber should have a water capacity of less than 2.5 litres/sq. metre of absorbing surface. (0.4 pints/sq.ft.). Even the slimmest radiators have a capacity which is double this, and several companies are marketing solar collectors based upon radiator panels which have a much higher capacity.

*4.2 Plastic Wafer Sandwich*
*4.3 Central Heating Radiator Type.*
*4.4 Roll-Bond Aluminium Plate.*

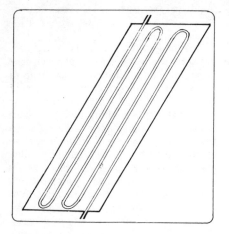

*4.5 Serpentine or Zig-Zag Tube-and-Sheet Absorber.*

## Tube-and-Sheet Absorbers

With tube-and-sheet absorbers, care must be exercised in the choice materials, the spacing of the tubes and the bond between the tube and sheet. The tube configuration may also have bearing upon the type of system in which the absorber can be used. A serpentine zig-zag configuration, for example, may offer excessive resistance to flow in a gravity circulation system unless it is so arranged that none of the bends force the circulation downwards. On the other hand a grid of parallel pipes between two header pipes might give problems due to air blockages if used in a drain-down system.

Tube-and-sheet absorbers, like the trickle type, have to transfer heat laterally, conducting energy collected on the absorbing sheet across to the water filled tubes. For this reason only good thermal conductors can be used. Unfortunately there seems to be some connection between high thermal conductivity and cost. The best heat conductors are materials like gold and silver. Of the commonly available materials copper is the best. It is almost twice as good as aluminium and nearly eight times better than mild steel. Copper is also the most expensive of the three. The ability of aluminium and steel to conduct the absorbed heat to the tubes can be improved by bringing the tubes closer together and by using thicker sheets. The thicker the sheet, the less the resistance to heat flow across it. Typical spacings and thicknesses for the different metals are as shown:

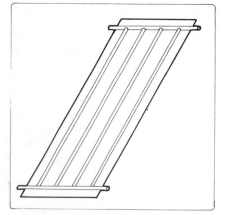

*4.6 Tube-and-sheet Absorber with Parallel Grid Configuration.*

| material | thickness | | spacing | |
|---|---|---|---|---|
| | mm. | SWG | mm. | ins |
| copper | 0.25 | 33 | 138 | 5½ |
| aluminium | 0.5 | 25 | 138 | 5½ |
| steel | 1.0 | 19 | 100 | 4 |

When seeking to make a good thermal bond between the tube and the sheet, the first step is to maximise the contact surface area. This can be done by forming the sheet round the tube or by deforming the tube to create a flatter contact surface. If the absorber is all copper, the tube and sheet should be soldered together. With other materials, some form of mechanical bond, clamping or clipping, perhaps in conjunction with a thermal paste, is necessary. Clamping systems should be examined to ensure that they do provide a tight contact along the whole length of the tube.

## CONDUCTION THROUGH THE CASING

If the case, or any part of it is metallic, it should not be allowed to come in direct contact with the absorber plate or its connecting pipework. This would form a 'heat bridge' enabling collected energy to leak away. Heat loss by conduction is reduced by insulating the case. The insulant must be able to withstand temperatures which may occasionally exceed 180°C. This rules out the use of polystyrene (styrofoam) which can melt at 85°C. 50-75 mm. (2 - 3 ins.) of a material like mineral wool or glass fibre quilt is adequate in single glazed collectors. Thinner layers will suffice if a material with a higher insulative value such as poly-urethane, fibre-glass sheeting or urea-formaldehyde, is used.

## CONVECTION CURRENTS

Convection currents around the absorber plate can be reduced by ensuring that the air gap between the plate and the cover is small — no more than 25mm (1 inch).

The cover itself of course greatly reduces heat losses simply by shielding the absorber from the cooling effect of wind. Convection losses can be further decreased by ensuring that the insulation on the underside is pressed up against the plate so as to prevent any air movement on the underside. Finally the case should be so designed as to have no air gaps which would allow through circulation of cold air. In practice though it is usually necessary, except in a few factory glazed and sealed collectors, to provide small vent and drain holes for the removal of condensation.

## RE-RADIATION LOSSES

The hotter a black body becomes, the more energy it will radiate. There are only two simple methods of reducing the radiative losses; through exploitation of the greenhouse effect and keeping operating temperatures to their lowest useful value. These have been described in the previous chapter.

## HIGH EFFICIENCY COLLECTORS

Various sophisticated techniques have been adopted to further decrease the heat losses from solar absorbers. These have in common a reliance upon industrial production methods difficult to reproduce in a home workshop, and considerable increases in cost. Bearing in mind the latter, it is worth noting that their effectiveness is most noticeable when operating at temperatures higher than is generally necessary for domestic water heating.

## Honeycombs

Honeycombs of transparent tubes have been mounted perpendicularly over the absorber plate. This suppresses convection and, if the diameter of the tube is small in proportion to its height, it will reduce re-radiation losses. Care must be taken that the honeycomb is not in contact with the absorber and the cover, but rather suspended between them. Otherwise conduction losses, and perhaps re-radiation, will increase.

## Selective Surfaces

Selective surfaces are extremely thin coatings electro-plated on to the absorber plate. They have the distinction that whilst absorbing a large proportion of solar radiation (not so much though, as a good matt black paint) the proportion of heat they re-radiate is very low. Their performance is often characterised by the ratio absorptivity : emissivity. An ordinary matt black surface might have a ratio of 90:90. whereas a good selective surface could have a ratio of 85:10. The emission factor is much lower and less heat is therefore radiated. This characteristic is very useful when operating at high temperatures. The selective surface is achieved by first applying a shiny coating such as bright nickel. This is then covered by a black absorbing layer such as black nickel. The thickness of this layer must be controlled so as to be only a few microns thick, i.e. less than the wavelength of the re-radiated heat waves.

The chemical structure of selective surfaces is such that they tend to be unstable in combination with copper and aluminium. There is therefore some doubt over just how many years they will continue to perform selectively at the values quoted for a new coating. For this reason there may be increasing interest in selectively coated glass. Coatings of tin oxide and indium oxide on the inner surface of a glass cover give results similar to that of a selective coating on the absorber. Such coatings have been used in other applications for many years without deterioration.

Selective coatings can usually be recognised by the Newton's rings, or rainbow coloured hue they reflect.

## Evacuated Collectors

If the surroundings of the absorber plate can be partially evacuated, the convective and conductive losses will be greatly reduced. The absorber will retain more of its heat, just like coffee in a thermos flask. Evacuated glass tubes with a selectively coated waterway running through their centre are the most efficient solar collectors available.

## Applications

High efficiency collectors will prove most useful in situations where high temperatures are required, e.g. solar cooling appliances, certain industrial processes, and any solar installation required to produce mechanical movement such as solar powered pumps. They will also be of use where the system is required to work primarily in difficult conditions e.g. space heating systems whose main load comes in the winter. Finally, a point which has not been given so much attention, because of their higher efficiency a much smaller area of these collectors is required when compared to conventional collectors — as little as half the standard recommended area in some cases. This means that they will prove useful in retrofitting i.e. the application of solar system to existing buildings. When dealing with buildings which are already standing, overshadowing and awkwardly shaped roofs sometimes restrict the area available for mounting collectors.

## LIFE EXPECTANCY

Having paid attention to all the physical aspects affecting the collector's efficiency, there remains a less obvious quality which a good solar must possess — longevity. Even a home made solar system will require several years to contribute heat to the value of the materials expended in its construction. Unless its life exceeds the duration of this payback period, the resources devoted to its installation could have been better used.

Defects in a solar collector can come from within or without; from the corrosive effect of the circulating fluid acting on the absorber plate, or from weathering of the case and glazing.

### Internal Corrosion

The combined presence of air and water always spells danger for thin sheets metal. Collectors in drain-down systems and direct systems are therefore most susceptible to attack and only strongly resistant materials such as copper and stainless steel should be used for the waterways in such systems. Pressed steel radiators should only be used in indirect systems where the quantity of air dissolved in the circulating water is limited and cannot be replenished.

A major cause of corrosion is mixing of metals in the solar circuit. Different metals have different electro-chemical charges and because of this, when two different metals are immersed in water, there will be an increased tendency for one of them to dissolve. This is particularly noticeable where water flows from a copper pipe into aluminium or galvanised iron.

Manufacturers of aluminium absorber plates stress that they should only be used in indirect systems with anti-corrosive additives in the water and that they should be mounted so as to be electrically isolated from any other metallic elements. Furthermore they should never be connected directly to copper pipes. A rod of magnesium is sometimes used, suspended as a sacrificial annode in the circulating fluid to 'use up' the corrosive ingredients in the water. It has been suggested the corrosion problems encountered with aluminium can be overcome by using oil instead of water as the circulating fluid which transfers heat from the collector to the store. Even with carefully selected oils however corrosion has occurred due to air becoming dissolved in the oil. Unless all the outlined precautions can be reliably maintained for the life of the installation, absorbers with aluminium waterways should be avoided.

Aluminium however can be useful when used as an absorbing sheet bonded to copper or stainless steel pipes in a tube-and-sheet absorber. In such designs, the joint between the two metals must be protectively coated in case moisture, condensation for example, should ever seep between them.

The plastics used in solar absorber construction are usually resistant to the corrosive effects of water. Hot water however can cause softening of most plastics and in some cases it can also leach out the stabilisers in the plastic. The ultra-violet component of solar radiation will also cause deterioration of many plastics. It is likely that suitable plastic absorbers will be developed in the near future, but for the present, it is safest to use them only in situations where the operating temperatures will be low, as in swimming pool water heating.

### External Weathering

The casing which supports and protects the absorber, and keeps the insulation dry, obviously has considerable influence in determining the life of the collector. Aluminium boxes with welded corners appear to be the most robust casings available. Care should be taken during their installation that they will not receive rainwater run-off from any copper pipes. Similarly, any connecting pipes entering the case should be physically separated from the aluminium by the use of gromets or sleeves made from a durable material such as neoprene. Cases moulded from fibre-glass also provide sound protection for the absorber. If the casing has been manufactured from a less durable material such as timber, it will have to have regular

attention to maintain its protective coating. Resinous timbers should be avoided as there is a danger that the resin will evaporate at high temperatures and condense on the underside of the glass thus reducing its transmissivity.

A weak point in many collectors is the joint between the glazing and the case. Conventional putty joints for example will have a very short life under direct sun and high temperatures if its surface is not regularly painted. Modern materials such as non-setting compounds, synthetic rubber, silicone sealants and pre-formed butyl strips are preferable alternatives having life-expectancies of 10-20 years.

### CONCLUSION

The solar collector is a relatively new item of hardware in plumbing systems. As such, there is as yet no commonly agreed method of testing their performance and presenting the results of such tests. It is therefore difficult to compare two different collectors when presented only with the manufacturer's data. The novelty of solar collectors has also led to some confusion among the general public regarding their function, and to some mystification of their operation on the part of some salesmen. They are in fact quite simple heat exchangers, absorbing solar radiation and transferring it to water. If they are constructed with regard to the points outlined in this chapter there is no reason why they should not carry out this function satisfactorily for many, many years.

# 5

# Circulation

It is by circulating water through the solar collector that the absorbed energy is brought to the storage tank, or directly to the point where it will be used. Water can be moved round the system in one of two ways: gravity circulation, also known as thermo-syphoning, and forced circulation achieved by means of a small pump or circulator.

As explained in chapter 7, the decision to use or not to use a pump is one of the first steps in planning a solar system. A pump will give much more freedom when it comes to locating the system components. Indeed, in some houses it will be impossible to circulate without a pump, because of the need to locate the storage tank above the collectors when relying on gravity circulation. A pumped system will also collect more energy than a comparable thermo-syphoning system. This is because it will achieve higher flow rates which will mean lower operating temperatures in the collector and, as is explained in Chapter 3, this will lead to higher efficiency. Another way of explaining the improved performance achieved with a pump is that it enables one to maintain

a higher rate of heat removal from the collector, thereby allowing the collector to absorb heat at a higher rate.

## Cost Penalties with Pumped Systems

The disadvantages of opting for pumped circulation are increased costs and the introduction of reliance on an electrical supply, without which the system will not operate. A suitable pump will cost between £15 and £25, wiring installation, about £4, the annual electrical bill will be about £5 and the best type of automatic pump switch controls sell for £35-£45. Against this can be set the savings accrued by using the smaller bore piping which is suitable for pumped systems.

## Gravity Circulation

The need for an electricity supply is unlikely to be a problem in most cases. There is however an undeniable attraction about the independence and self

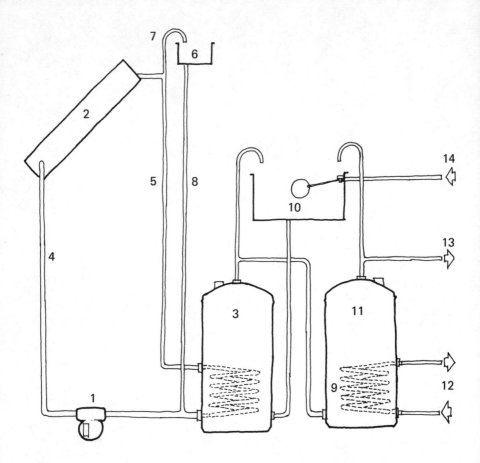

## 5.1 Pumped System — Typical Layout

1. Circulator or pump.
2. Solar Collector.
3. Solar Storage Tank.
4. Return pipe carrying cooled water to collector.
5. Flow pipe carrying heated water to solar tank.
6. Solar primary circuit expansion tank.

7. Solar primary vent pipe.
8. Solar primary feed pipe.
9. Heat exchanger.
10. Cold feed tank.
11. Hot water cylinder.
12. Connections to a boiler.
13. To hot water taps.
14. Mains water supply.

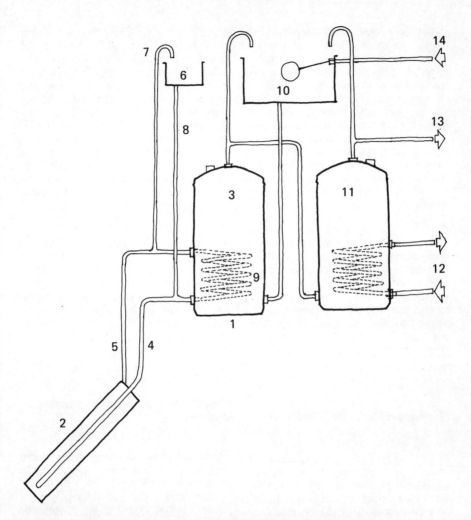

## 5.2 Gravity Circulation (Thermo-syphoning) System — Typical Layout

1. Bottom of solar storage tank which must be higher than top of solar collector.
2. Solar Collector.
3. Solar Storage Tank.
4. Return pipe carrying cooled water to collector.
5. Flow pipe carrying heated water to solar tank.
6. Solar primary circuit expansion tank.
7. Solar primary vent pipe.
8. Solar primary feed pipe.
9. Heat exchanger.
10. Cold feed tank.
11. Hot water cylinder.
12. Connections to a boiler.
13. To hot water taps.
14. Mains water supply.

regulating nature of the gravity circulation systems. In these, the storage is located above the collectors so that as water in the collectors becomes heated, its tendency to rise in a convection current, brings it up to the storage tank, at the same time drawing cold water from the bottom of the storage tank into the bottom of the collectors. The stronger the radiation arriving on the collector, the faster the circulation. And, of course, when there is no energy to collect, circulation will cease automatically.

The principle of thermo-syphonage as utilised in gravity circulation systems is therefore very simple. To ensure that it works well in practice, some care must be taken in pipe layout, pipe sizing, storage location and the collector connection pattern. All these factors effect the resistance to flow through the circuit and this must be minimised.

## Pipe Layout for Gravity Circulation

Resistance to flow through pipes can be reduced by utilising larger diameter pipes, and minimising the length of pipe and the number of bends and fittings. The convective current of heated water can also be facilitated by ensuring that, between the collectors and the storage tank, there is a continuous rising gradient in the pipework.

Similarly, there should be a continuous fall in the piping connecting the bottom of the storage tank to the bottom of the collectors.

In the horizontal plane, the collectors and tank should be as close as possible

to each other, in order that the length of pipework be small. In the vertical plane, increasing height of the tank above the collectors by up to about 4 metres (40 feet) can improve the thermo-syphonic flow. Greater separation will cause a drop in circulation rate. The advantages of a large vertical separation between collectors and store might be outweighed however if this also meant longer runs in the distribution pipework between the storage tank and the points of use, as this would result in increased heat losses.

In all cases, the storage tank should be at least 600 mm (2 feet) above the top of the collectors to ensure that there will be no back syphonage on cold nights. If this occurred, warm water from the storage tank might be drawn into the collectors where it would be cooled by radiating its heat to the night sky.

## Pipe Diameter in Gravity Circulation Systems

Although I have produced tanks full of hot water on very sunny days using 15 mm diameter (½ inch) pipe, I would not recommend anyone to use pipe smaller than 28 mm (1 inch) for a domestic installation. Larger diameter copper pipe is very expensive, and as explained in chapter 8, a bending machine is required for pipes larger than 22 mm (¾ inch). Because of this, plastic pipes have a special attractiveness for gravity circulation systems. Thick walled polythene pipe is available in large diameters and bends with low resistance to flow can easily be made by flexing the pipe in a gradual sweep. When using plastic pipe it must be supported at regular intervals to prevent it sagging when loaded with hot water. Such sagging can oppose the gradual rise or fall which is required in pipework for thermo-syphoning.

The table in appendix 4 should be of help in choosing the optimum sizes of pipes for particular systems.

## Gravity Circulation and Solar Collectors

As shown in the last chapter, there is a considerable variety in the construction and configuration of solar collectors. Some are better than others for facilitating a thermo-syphonic flow.

The open trough type, trickle collector and coil configurations cannot be used in a gravity circulation system as they will not function without a pump.

Tube on strip collectors where the tube zig-zags across the plate in a serpentine configuration can offer considerable resistance to water flow. The best ones are those which use at least 22 mm (¾ inch) pipe and where there is an obvious gradient in the horizontal pipe runs.

Sandwich type collectors, such as radiator panels, will offer comparative little resistance to flow. In the case of radiators however, their water capacity and hence their thermal mass, is high. As explained in chapter 3, this will mean that the collector will not effectively respond to short outbursts of sun. If this type of collector is used, it should be mounted so that any ribs or reinforcement which form channels inside the absorber, run vertically so as not to obstruct the upward flow of warm water. It also helps if they are mounted at a slight tilt so that the horizontal headers have a slight gradient.

Tube on strip collectors which have a grid layout of parallel tubes between two horizontal header pipes are the best choice for a gravity circulation system.

Because of the slower reaction of thermo-syphoning systems, and their higher operating temperatures, double glazing and increased insulation may be advantageous in the collectors.

## Choice of Pump

If circulation is to be forced, a suitable pump must be chosen. It should be electrically rated at a minimum of 20 watts. It should also have a guarantee to withstand corrosion and temperatures up to 100°C. The circulators usually used for domestic central heating are the type to look for. These are easily available, quiet running, and have a power consumption of about 100W. They have traditionally been produced in a variety of ratings to suit different applications. The rating reflects the resistance to flow which it can overcome, and this is expressed in terms of the head of water, which would exert a force equivalent to this resistance. Typically a pump with a 3.5 metre rating would be suitable for a system where the storage was in the attic, close to the collectors. Where the pipe runs linking the store and collectors are long, a 4.5 metre head pump might be necessary. 15 mm (½ inch) pipe is usually used in pumped systems. If smaller diameter, microbore pipe, is used, the additional resistance which this will cause must be overcome by using a pump with a 5.5 metre head.

Choice of pump rating can be simplified by choosing a variable head pump. These are now produced by the major manufacturers at the same price as the single rated pumps. The pump which I am most familiar with, the Grundfoss model, also has a two way speed selector switch. When installed the pump should be set at its lowest rating and speed. If the flow is too slow, as indicated by high temperatures between the collector inlet and outlet the speed can be increased. If this is still inadequate, the rating selector can be changed.

**Pumps in Drain down Systems**

If the collectors are to be drained down when there is insufficient energy to warrant circulation, a submersible pump inside an open top storage tank is the most appropriate. Other pumps can be used but care must be taken in locating them well below the level of water in the storage tank. Most pumps are not self priming and if there is not water in the pipework on each side of the pump when it is turned on, it is likely to burn out its motor.

**Pumps in Direct Systems**

Direct systems suffer much more from the effects of corrosion since fresh aerated water is continually being introduced. Only bronze pumps should be used in direct systems, as these are less liable to corrode.

**Control Switch for Pump**

As sure as night follows day, there will

not always be enough energy being collected to warrant running the pump continuously. A control mechanism is required.

In sunny climates, a manual switch or a simple time switch will suffice, turning the pump on in the morning, and off in the evening or mid-afternoon. In cloudy countries where the pattern of sunshine is less predictable, a system left running all day will frequently be radiating heat to the sky instead of vice versa. Unfortunately it will not improve the weather, and it certainly will not decrease fuel bills.

The pump could be controlled by a radiation sensitive device such as a solar cell which would switch the pump on whenever solar radiation exceeded a preselected level, say 150 watts per square metre. This could also be done with an ordinary room thermostat mounted on the back of a small sheet of blackened aluminium set inside an insulated glazed box, a mini solar collector, mounted beside the real ones. The thermostat should be given a low temperature setting, but remember that the sheet of aluminium will heat up much quicker than the water filled panels. The thermostat should be kept accessible so that adjustments can be made to find the best setting for any particular situation.

Such radiation sensitive controls can be made without spending a lot of money. There will however be occasions when the system will be throwing heat away, e.g. after a bright sunny morning the storage tank might be full of very hot water. If the afternoon is only moderately bright, the radiation sensor is still likely to keep the pump running and hence lower the tank temperature.

The most accurate type of pump controller for solar systems is the one known as a differential thermostatic controller, otherwise called a 'black-box'. The box is fed by four sets of wires; one

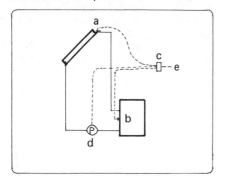

5.3 Diagram of Pump Controller Layout. a) temperature sensor mounted on solar collector; b) sensor on storage tank; c) 'black-box' control; d) pump; e) electricity supply.

brings power from the mains; a second takes power to the pump; the remaining two are connections to temperature sensors, one attached to the storage tank, the other attached to the collector outlet. Inside the box, a simple electronic circuit compares the two temperature readings. Only when the collector is warmer than the water in the storage tank will the pump be switched on.

The electronic circuit is sometimes made a little more sophisitcated by adding a selector switch which allows the installer to choose by how many degrees the collector should be warmer than the tank before starting circulation. The usual setting is 3-5 degrees C. This difference is to ensure that the energy collected is greater than the amount of energy required to run the pump.

A problem frequently encountered with control systems is that known as 'cycling'.

When this happens, the pump continues to switch on and off in very short time cycles. The problem here is that water in the collector becomes heated and the differential controller switches the pump on. This brings an immediate rush of fresh cold water into the collector which cools it down thus causing the controller to turn the pump off. This becomes a serious problem in systems with long pipe runs. Warm water from the collectors would be forced out of the collectors but may never reach the tank before the pump is switched off. This would mean that the heated water would lay in the flow pipe and would perhaps have cooled down before the collectors became hot enough to cause the pump to switch on again.

Some controllers counter this problem by including a time lapse device which will prevent the controller from turning the pump off until a predetermined interval after the collector is cooled. The time interval required is equal to the time taken for water to flow from the collector inlet to the tank inlet. This period can be measured by leaving the pump switched off on a sunny morning so that the pipework is cold and the water in the collector very hot. With one hand on a stop watch and the other on the tank inlet, one can measure the time between switching on the pump manually and feeling the pipe heating up. This indicates the time water takes to flow to the tank from the collector outlet. After a few seconds the temperature will fall again, indicating that all the water that was laying in the collectors has been pumped through to the tank. This is the time interval which should be registered on the controller.

One of the most accurate controllers on the market, that manufactured by Air Distribution Ltd., has a different modification. It has a temperature related time lag which will allow the pump to continue running for up to 3 minutes if it is only ½°C cooler than the tank. The greater the cooling effect of the solar collectors, the sooner the pump will be cut off.

## Fixing the Thermal Sensors

The thermal sensors, usually thermistors, upon which the differential temperature controller relies, must have a good thermal contact with the collector outlet and the tank. They can be fixed with an epoxy resin, tucked into a tight fitting copper sleeve which has been previously soldered onto the pipe, or strapped on firmly with a non-setting thermal paste to aid thermal contact. They should both be well insulated after fixing. The most accurate responses come from sensors immersed in the water.

The collector sensor should be situated immediately outside, or even inside the collector casing. The storage tank sensor should be situated mid way between the inlet and outlet leading to the collectors.

## Freeze Protection

Due to the exposed position of solar collectors, there is a danger that the circulating fluid might freeze during cold nights. The danger is greatest on nights with clear skies as the collector temperature can drop below the surrounding air temperature due to radiation of heat into space.

As water approaches its freezing point, it expands in volume. This expansion can fracture the metal waterways causing them to leak when the liquid defrosts. Steps must be taken therefore to pre-

vent freezing of the fluid in solar collectors. There are five possible solutions to this;

i  Addition of anti-freeze solution to the circulating fluid.
ii  Draindown of collectors.
iii  Heating the collectors, either by circulating warm water from the tank or by an electrical resistance input.
iv  Insertion of compressible gas-filled tubes inside the waterways to accept the expansion experienced on freezing.
v  Insulated covers.

## Anti-Freeze

The addition of an anti-freeze solution is the most commonly adopted freeze protection method. It does have disadvantages. It necessitates the use of a heat exchanger in the storage tank. Suitable anti-freeze solutions are expensive and even these will need replacing after 5—10 years.

The anti-freezes usually advocated for solar systems are compounds based on ethylene-glycol. This has risen considerably in price in recent years and cheaper alternatives based upon methanol have been introduced. These should never be used however due to the toxicity of the vapour which they give off when heated. Concern has also been expressed over the toxicity of ethylene-glycol. It will only be dangerous however if it leaks through the heat exchanger and into the hot water tank. Even then, it will only be harmful if it is drunk in very large quantities. On account of this slight danger, propylene glycol has been used. It is however, still more expensive.

One of the most suitable commercially

prepared anti-freeze solutions is Fernox FP—1. This is an ethylene glycol based compound but it contains a taste spoiler and a foaming agent which will act as signals should the anti-freeze ever find its way into the tap water. It also contains corrosion inhibitors which would counter the danger of glycollic acid formation.

The manufacturers of Fernox FP—1 also point out that leakage of anti-freeze into the secondary water circuit can be prevented simply by ensuring that the water level in the solar primary circuit's expansion tank is below the water level in the secondary circuit's cold feed tank. Given this arrangement, even if the heat exchanger does leak, the higher pressure in the secondary circuit will cause the tap water to leak into the primary circuit and overflow the expansion tank rather than a leakage in the opposite direction. If an overflow pipe has been fitted to the expansion tank to discharge over the roof, constant discharge from this pipe will signal a fault.

### How much Anti-freeze?

The quantity of anti-freeze required depends upon the water capacity of the primary circuit pipework and the collectors, and upon the physical properties of the solution in reducing the freezing point. A typical domestic system will hold about 36 litres (8 gallons) in the primary circuit and about 20%—25% of this should be anti-freeze if ethylene glycol is used. (i.e. up to 9 litres or 2 gallons).

If you wish to calculate the exact quantities for your own system, you will have to check the trade literature for water content of the particular type of collectors you are using, and measure the length of pipework, not forgetting the heat exchanger coil. Knowing the length of pipework, the following table will help you calculate its water capacity.

| Tube Diameter | Water Contents | |
|---|---|---|
| 10 mm | 0.06 | litres/metre |
| 15 mm | 0.15 | litres/metre |
| 22 mm | 0.32 | litres/metre |
| 28 mm | 0.54 | litres/metre |

The manufacturers of Fernox FP—1 recommend using at least 10% of their antifreeze in the circulating water. Stronger solutions will provide lower freezing points as indicated in the table below:

| Solution | Approx Freezing Point | |
|---|---|---|
| plain water | 0°C | 32°F |
| plain water with 10% FP—1 | −4 | 25 |
| plain water with 15% FP—1 | −7 | 20 |
| plain water with 20% FP—1 | −10 | 14 |
| plain water with 25% FP—1 | −14 | 7 |
| plain water with 30% FP—1 | −18 | 0 |

### Heating Collectors at Night

A simple thermostat in the collectors could start circulation of warm water from the storage tank, or switch on electrical resistance heaters inside the panels whenever their temperature approached freezing point. This at first seems a terrible waste of energy but in many situations the energy gains achieved by operating without a heat exchanger can be greater than the energy expended on heating on cold nights.

## Compressible Tubes

The insertion of gas-filled compressible tubes inside the collectors waterways is an attractive method of dealing with the freezing problem. There are of course practical difficulties and as yet the technique has not been widely applied.

## Insulated Covers

Covers which fit over the face of the collectors and prevent freezing at night are probably the simplest and cheapest solution. Unfortunately it depends upon the frailty of human memory unless automatic blinds are introduced and that would be very expensive. The risk of damage if the collectors were left uncovered is probably too great to make covers a practical solution in all but a few cases.

## Draindown of Collectors

Draindown systems are more liable to corrosion due to the continual introduction of air into the wet inner surfaces of the collectors. Special attention is also needed to prevent air blockages occurring. Copper and stainless steel should be able to withstand corrosion attacks for many years, but a more difficult problem is the formation of scale in regions with hard water supplies. This deposit on the inside of the collectors reduces the transfer of heat between the absorber and the circulating water. It would have to be chemically removed every few years.

# 6

# Storage

The heat store is the root of the solar system. Without it hot water would be available only when the sun is actually shining. A storage tank allows the solar system to operate whenever energy is available and to supply heat when it is needed.

## Size

The size of the storage tank will be determined by the size of the collectors and the amount of hot water required in the course of a day. An easily remembered rule of thumb is ... provide 1 gallon of storage for every 1 sq.ft. of solar collector. In practise, this might be increased to as much as 1.3 galls per sq. ft. (The metric equivalent would be 50-63 litres per square metre of collector). Although one could increase the total collected by increasing the storage capacity still further, the temperature reached in the tank would be much lower unless special precautions were taken to encourage temperature stratification.

## Choice of Tank

The usual choice for a storage tank is a copper cylinder as used for hot water storage in conventional heating systems. These have many advantages including flexibility in location, ease of installation, reduced evaporation losses, and longevity. But they are one of the most expensive items on the shopping list for a D.I.Y. system.

Galvanised iron, Polypropylene, stainless steel and glass fibre have been suggested as alternatives to copper cylinders. Galvanised tanks are liable to corrosion when linked in circuits with copper. Stainless steel and glass fibre cylinders are no cheaper than copper at present. Only polypropylene and glass fibre open top tanks therefore are worth considering and care should be taken in the choice of glass fibre tanks as not all glass fibre resins are suitable for high temperatures.

## Heat Loss from Open Top Tanks

The heat loss in open top tanks is greater than in cylinders because of evaporation which occurs across the air water surface. Furthermore the drop in the water level which occurs during draw-offs tend to draw in cold outside air. This absorbs heat which is then expelled as the tank refills. This will occur even when a tight fitting lid is used. The solution to these heat loss problems is to use a floating layer of insulation. Polypropylene balls are manufactured for similar industrial applications

A more easily obtainable material would be polystyrene granules. This material might melt however if the water temperature exceeded 65°C. The insulating balls should be of a diameter larger than the 22 mm (¾ inch) draw-off pipe to prevent blockages occurring. A cover is still required to keep impurities out of the tank.

## Cold Feed to an Open Top Tank

The open top tank could be filled via a ball cock from the mains supply as in a cold water storage tank. This would tend to disrupt the stratification in the tank mixing cold mains water with the hottest water which would have risen to the top of the tank. This problem might have been avoided by using a silencer pipe which carried the cold mains water to the bottom of the tank. But nowadays these are forbidden by Waterboard Regulations in order to prevent any danger of contamination from the tank passing into the mains pipe via the silencer tube. Furthermore most Water Authorities forbid mains connections to hot water tanks of more than a few litres capacity. An alternative to direct feeding with a ball valve, is a separate feed tank as would be used with a cylinder. With an open tank, however, the relative locations of the cold feed tank and the hot storage must be carefully selected as the water levels in each will be the same. Whilst a cylinder can be positioned anywhere below the feed tank, an open top tank must be placed level with it, otherwise it will overflow. This requires more floor area and open top tanks are therefore usually mounted in attics, beside the existing cold water storage tank which can then serve as its cold feed.

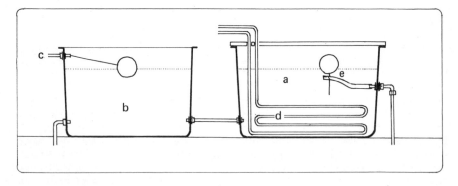

6.1 a) Open-top storage tank level with b) cold storage/feed tank c) mains water supply via ballcock d) heat exchanger e) warm water draw-off via flexible plastic tube connected to a float to ensure that draw-off is taken from the uppermost layer of water.

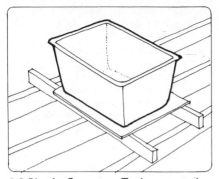

6.2 *Plastic Open-top Tank mounted on a flat even surface with weight spread on bearers.*

If a new cold feed tank is required, this should be an open top PVC tank with a capacity equal to about half the capacity of the hot storage tank with a 28 mm or, at least 22 mm (1 inch or ¾ inch) connecting pipe. A smaller tank might serve as well but there is a possibility that, when drawing off large quantities of hot water, as at bath times, the ball cock will not refill the system quickly enough to prevent the water level in the hot tank dropping below the draw-off outlet and thereby letting air into the pipework with resultant hissing and spurting at the hot taps.

The cold feed tank is always the highest point in a plumbing system, and any vent pipes from cylinders connected directly to it, should be installed to discharge over it.

**Structural Considerations**

It must be remembered when installing water tanks in the attic that a tank full of water is very heavy. Each litre weighs 1 kg. (1 cubic foot weighs about 10 lbs). A family size 200 litre (44 gallon) hot water tank will weigh 200 kg (almost 4 cwt.) Tanks should therefore be located over load bearing walls where possible and their weight should be spread over several joists by the use of bearers.

6.3 *Galvanized Iron Open-top Tank mounted directly on bearers.*

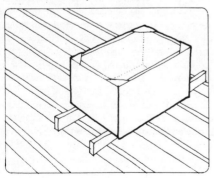

Plastic tanks require support across the whole of their base surface. A suitably sized sheet of ply or chipboard should therefore be laid over the supporting bearers. With galvanised iron tanks, such practice might lead to condensation collecting on the underside of the tank and subsequent corrosion of the tank. They should always be mounted directly on their bearers.

41

In chapter 7 the choice of direct or indirect systems is discussed. In indirect systems, a heat exchanger will be required in the storage tank. This serves to separate the water flowing through the collectors from the water which flows to the taps inside the house. It must of course allow the heat absorbed by the water in the collectors to pass into the water in the storage tank.

No heat exchanger can transfer all the heat absorbed by the primary circuit into the secondary circuit. Normally only 60–90% is exchanged. A considerable proportion of the solar energy collected, therefore, may never reach the storage tank because of the inefficiency of heat exchangers. It is worthwhile therefore looking at the factors which affect their performance.

The amount of heat transferred across a heat exchanger depends upon four conditions:
  (i) The temperature difference between the hot water flowing into the exchanger and the water to be heated in the tank.
  (ii) The thermal conductivity of the heat exchanger i.e. the ease with which heat can flow through the walls of the heat exchanger. Metals generally have a high thermal conductivity and copper is particularly good.
  (iii) The surface area of the exchanger. The larger the area of interface between the hot and cold water, the more heat can flow between them.
  (iv) The water flow rate. The more water that flows through the exchanger the greater will be the quantity of heat that passes through it.

The heat exchanger should be positioned in the bottom of the solar storage tank where the coldest water settles. This is where the greatest temperature difference can be found, and hence where the exchanger will operate most effectively. This is extremely important in solar systems as the hot water coming from the solar collectors will on many occasions be only a few degrees above the temperature of the water in the tank.

*6.4 Heat exchange coil in the bottom of a hot water cylinder.*

**Heat Exchanger Material**

The best material for constructing a heat exchanger is copper. Plastics are generally insulative and tend to resist the through flow of heat. A cheap solution which has been suggested by recycling enthusiasts is the use of heat exchangers salvaged from old refrigerators or motor car radiators. These may prove effective for a short period but will undoubtedly run into problems with corrosion after a few months.

## Heat Exchanger Surface Area

In a domestic-size installation the surface area of the heat exchanger should be about 0.2-0.3 square metres per square metre of collector surface. The simplest form is a long length of pipe immersed in the storage tank. For each square metre of collector there should be 4 metres length of 15 mm pipe or 3 metres length of 22 mm pipe as heat exchanger. If space inside the tank is restricted, finned pipe can be used. This provides an increased surface area for a comparable length of pipe.

## Heat Exchanger Geometry

The geometric form of the heat exchanger will effect the rate of flow of the water passing through it. Sharp bends increase the resistance to flow, slow it down, and thus decrease the rate of heat transfer. Gradual "slow" bends are therefore preferable to sharp elbows. This consideration is of particular importance in thermo-syphoning systems where the flow is already much slower. In such cases a grid of pipes, as shown in figure 6.5 will present less resistance to flow than a long length of pipe with many bends.

In pumped systems though, a much simpler exchanger can be made. Using a plumbers bending spring as explained in chapter 8, form an appropriate length of 15 mm diameter pipe into a serpentine, zig-zag arrangement. This should be dimensioned to fit into the bottom of the tank walls. It should also be supported inside the tank to raise it clear of the base, where it might become covered by sludge deposits. Bricks can serve as supports.

In pumped systems, where resistance to flow can easily be overcome, bends in the heat exchanger can be beneficial. They can increase the flow turbulence and hence the heat transfer. An even simpler solution would be to use microbore piping which is easily flexed into the desired shape. It is sold in coils which could simply be spaced apart and submerged in the tank. Allow 7 metres of 10 mm pipe for every square metre of collector (2 ft. length per sq. ft. of collector).

## Connecting the Exchanger in Open Top Tanks

With pumped systems the connecting pipes between solar collectors and heat exchangers can be simply dropped into

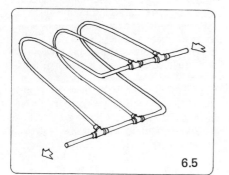

6.5

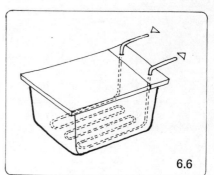

6.6

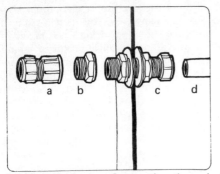

6.7 Heat Exchanger Connection through Tank Wall. a) straight couple; copper to female iron b) Reducing bush c) straight couple; copper to male iron with extended shank, two backnuts and washers d) copper heat exchanger.

the tank from above as in figure 6.6. It would be advisable to lag the outlet of the heat exchanger with some waterproof insulation where it passes vertically through the upper layer of water in the storage tank. Otherwise it will on occasion be taking heat away from the top of the tank.

In thermo-syphoning systems, dropping pipes from the top of the tank into the heat exchanger forms additional resistance to flow and it is therefore better to connect the exchanger through holes drilled in the walls of the tank, as shown in figure 6.7.

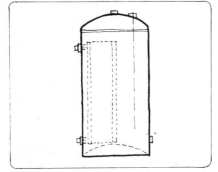

6.8 Indirect cylinder with an anulan heat exchanger – unsuitable as a solar tank.

6.9 Indirect cylinder with a coil heat exchanger.

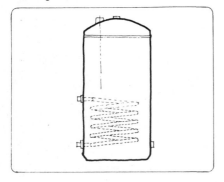

**Heat Exchangers in Cylinders**

Copper cylinders which incorporate heat exchangers are widely available. These are known as "indirect" cylinders. There are three main types: Primatic, anular and coil. Coil heat exchangers are the only type suitable for solar systems. Indirect cylinders are usually connected to gas boilers and hence operate with a big temperature difference. Because of this their surface area is unfortunately not so large as would be desirable in a solar storage tank. To ensure selecting the cylinder with the most appropriate heat exchanger when buying a new cylinder one should seek those approved under British Standard 1566. These will have "B.S. 1566" and a kite mark stamped on their sides.

**British Standard Approved Indirect Copper Cylinders (B.S. 1566)**

| Metric | | | | Imperial | |
|---|---|---|---|---|---|
| Dia. | Height | Capacity | | Dia. | Height |
| millimetres | | Litres/Gals | | Inches | |
| 300 | 1600 | 96 | 21 | 12 | 64 |
| 350 | 900 | 72 | 16 | 14 | 36 |
| 400 | 900 | 96 | 21 | 16 | 36 |
| 400 | 1050 | 114 | 25 | 16 | 42 |
| 450 | 675 | 84 | 18 | 18 | 27 |
| 450 | 750 | 95 | 21 | 18 | 30 |
| 450 | 825 | 106 | 23 | 18 | 33 |
| 450 | 900 | 117 | 26 | 18 | 38 |
| 450 | 1050 | 140 | 31 | 18 | 42 |
| 450 | 1200 | 162 | 36 | 18 | 48 |
| 500 | 1200 | 190 | 42 | 20 | 48 |
| 500 | 1500 | 245 | 54 | 20 | 60 |
| 600 | 1200 | 280 | 62 | 24 | 48 |
| 600 | 1500 | 360 | 79 | 24 | 60 |
| 600 | 1800 | 440 | 97 | 24 | 72 |

In addition to these, I.M.I. manufacture a specially designed solar cylinder which has a larger than normal heat exchanger.

If you find a second hand cylinder it is quite likely to be a "direct" model i.e. without a heat exchanger. Access to the interior of the cylinder is restricted and it is therefore difficult to install a home made heat exchanger. There are however proprietary brands available. If there is a 2¼" boss for an electric immersion heater on the tank, then a heat exchanger such as a "Hot Rod", or a "Micraversion" can be simply screwed into this boss. The immersion boss is situated either for vertical entry at the top of the cylinder, or for horizontal

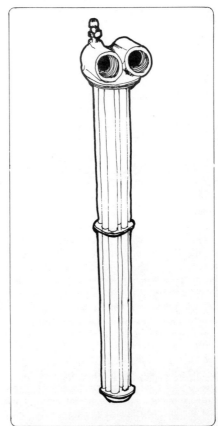

6.11 "Micraversion" heat exchanger. It can be used with gravity circulation due to parallel flow arrangement shown in fig.6.12 on next page.

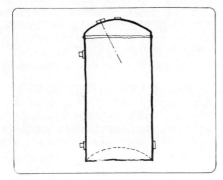

6.10 Direct cylinder with boss for a top-entry immersion heater.

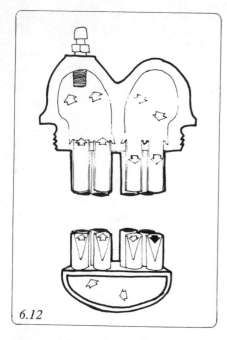

6.12

## Counter Flow Heat Exchangers

The most efficient heat exchangers are those where the fluids between which heat is to be transferred are made to flow in opposite directions on each side of a division which has a high thermal conductivity. These are placed outside the storage tank and an additional pump is required to circulate water from the storage to the heat exchanger. This additional expense makes the counter flow heat exchanger impractical for domestic installations. In larger-scale applications however, it is worth consideration.

entry at the bottom. As already discussed, the heat exchanger will operate more efficiently at the bottom of the tank. If you already have a tank with the boss at the top, it is worth while mounting the cylinder on its side with the boss on the underside, and the hot water draw and vent pipe being taken from an outlet on the upper side. The cylinder should be slightly inclined so that the outlet is at the highest point. Copper cylinders dent easily, so if you do decide to install one horizontally, use a sheet of ply or chipboard to provide an even support underneath, and use wedges to prevent rolling.

If only a direct cylinder without an immersion boss is available, it is still possible to fit a heat exchanger known as a "sidewinder". This is a bit more difficult to install as it involves cutting holes in the wall of the cylinder. The manufacturers instructions should be followed.

## Expansion Tanks

In addition to hot storage and cold feed tanks, a third small tank is necessary in an indirect solar system. This serves to take up the expansion in volume which occurs when water in the primary circuit, that which links the collectors to the heat exchanger in the hot storage tank, is heated. A 5 litre (about 1 gallon) polypropylene tank is ideal. This tank also serves as a filling point when commissioning the system, and collects any overflow which might come out of the primary circuit vent pipe which is positioned to discharge over it.

The expansion tank must always be the highest point in the primary circuit. It should also be borne in mind when installing it that space above and around it is required for access when filling, adding anti-freeze and checking the water level is being maintained every few months.

## Pressurized Vessels

The requirement that the expansion tank be at the top of the system sometimes presents space problems. In modern shallow tilted roofs there is often insufficient height to mount the tank above roof mounted collectors. In such cases

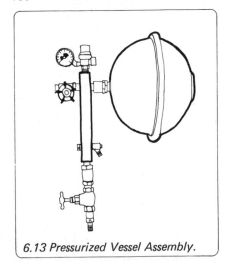

*6.13 Pressurized Vessel Assembly.*

a pressurized vessel must be used. This is quite simply an air tight tank divided in two by a flexible diaphragm. On one side a gas such as nitrogen is sealed in. The other side is open to water from the system to which it is attached. If this water expands in volume, the expansion will be taken up by a stretching of the diaphragm and a compression of the gas.

The benefit of the pressurized vessel is that it can be positioned any where in the system that is considered convenient. It has the additional advantages that there will be reduced evaporation losses from the primary circuit as compared with open top expansion tanks, and less air infiltration which should mean less possibility of corrosion in the collectors.

The disadvantage is the additional cost. Not only is the pressurized vessel more expensive than an open top plastic tank, but a special filling unit is required.

The system can be filled with water in one of two ways when a pressurized vessel is used—by gravity or by mains pressure. To fill by gravity, one must again have access to the highest point in the system. IMI solar systems use this method and incorporate a filler cap, as is used in car radiators, on a tee branch off the uppermost run of piping.

Mains pressure filling is the more usual method adopted when pressurized vessels are used in central heating systems. This requires a filling assembly which consists of the following: non-return valve, which ensures that water from the system which will contain anti freeze, cannot flow back to contaminate the mains supply; a stop cock, to close off the system after filling; a pressure gauge, to indicate any fall in pressure in the system, as would be caused by a leak; a pressure release valve, a safety device which prevents pressures inside the system becoming dangerously high; and sometimes a drain cock, for emptying the system. Most, if not all, Water Authorities disapprove of permanent mains connections to pressurized systems. It is necessary then for the filling assembly to terminate in a hose pipe connector which will facilitate a temporary connection to the mains. Sometimes an isolating valve is included so that the system when filled can be made independent of the fitting assembly.

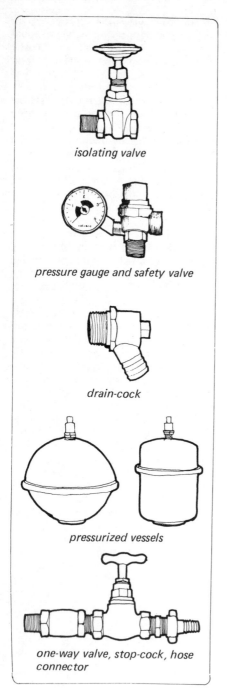

*isolating valve*

*pressure gauge and safety valve*

*drain-cock*

*pressurized vessels*

*one-way valve, stop-cock, hose connector*

In addition to the filling assembly, an air cock is required at the highest point in the system. In installation and filling the manufacturer's recommendations should be followed. The normal procedure after connecting the mains via a hose to the filling assembly is to open the stop cock until the pressure gauge registers about 1.5 atmospheres. During filling, air cock should be left open until water begins to leak through it. It should then be closed, hand tight. After the first sunny day when the primary circuit has been heated and the pump has been running, air bubbles are likely to have collected and should be released by opening the air cock. After this the pressure in the system may have dropped below 1.5 and will need to be replenished by opening the stop cock to reintroduce mains pressure.

The size of the pressurized vessel required depends upon the amount of expansion which is expected.

Water will expand by about 5% when heated from very cold to very hot. Hence 4 square metres of collector, the heat exchanger and the connecting pipework, which will probably contain no more than 40 litres (9 gallons) is liable to expand by about 2 litres (less than ½ gallon). Antifreeze however expands by a greater amount and it is wise to play safe and allow for double this amount.

**Space Saving Tanks**

Various units containing different combinations of tanks and heat exchangers are available. These not only require less space but usually require less time to install as many of the connections are pre-plumbed. These units do however

48

cost more than the sum total cost of the separate elements which they replace.

IMI manufacture an indirect copper cylinder which contains two coil heat exchangers. Solar collectors could be fitted to the lowest heat exchanger with a conventional boiler providing a temperature boost via the second exchanger in the middle of the tank. This would be suitable for use in the one-tank system described in chapter 7.

Several companies produce combination cylinders, either direct or indirect, like the "Fortec" model which is a single copper unit comprising an open top, cold storage tank which feeds into a cylinder below. These are also available in an oval shape which will fit into a shallow cupboard which would be too small for a conventional circular plan cylinder.

Larger combinations of tanks are also available in the form of "Harcopacks". These consist of a rectangular galvanised cold water tank supported on an

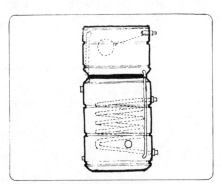

*6.15 Combination Cylinder*

angle iron frame, above a copper cylinder. A third vessel, a small expansion and primary feed tank can also be integrated in the unit.

## Insulation

Do not forget that even with the best tanks and heat exchangers, very little heat will be stored without the help of insulation. Tanks should have at least 100 mm (4 in.) of a good thermally insulating material wrapped all round them. Use extra at the top where the tank will be hottest.

Cylinders can be bought with a 25 mm (1 in.) layer of foam already applied. This should be further insulated with a cylinder jacket. An unfoamed cylinder should have additional glass fibre quilt wrapped around it underneath the jacket. Remember though that glass fibre quilt is an irritant to your skin. So if you are working in confined spaces, it might be worth the additional expense of buying two cylinder jackets. These jackets are simply plastic covered glass fibre made in a variety of sizes to fit the standard cylinder dimensions.

Rectangular tanks can be insulated in sheets of rigid polystyrene or fibre glass.

# 7

# Choosing a System

The solar collector and storage tank, and a means of circulation between them are present in all solar water heating systems suitable for temperate climates. There is, however, considerable variation in the systems of connecting them. The final choice is determined by four factors:

1) Cost
2) Available locations for collectors
3) Need for frost protection
4) Type of back-up system for temperature boosting.

The cheapest systems are obviously those requiring the minimum of equipment. The location and construction of the house may, however, necessitate the use of pumps and special controls. The use of a second tank, whilst adding to the cost, may facilitate plumbing work and save disturbing the existing heating installation.

In this chapter we look at the two basic systems — thermosyphoning and pumped —

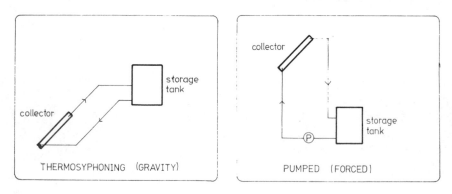

THERMOSYPHONING (GRAVITY)     PUMPED (FORCED)

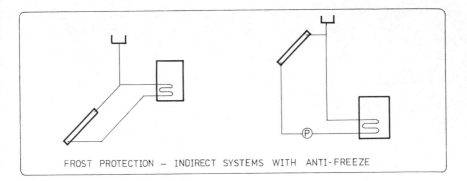

FROST PROTECTION — INDIRECT SYSTEMS WITH ANTI-FREEZE

and examine the possible variations of each. The first step is to decide between thermosyphoning and pumping. As explained in chapter 5, if you wish to use thermosyphoning circulation, you must be able to place the collectors below the storage tank and reasonably close to it.

Secondly, one must decide on a method of protecting the collector from damage by freezing. Both systems can be protected by adopting an indirect system. This means using a storage tank with a heat exchange coil in it.

One can then add an anti-freeze solution to the water passing through the collectors without contaminating the water in the storage tank. Note that a separate expansion tank is required for the "primary" loop carrying water through the collectors. This must be the highest point in the system unless a pressurized tank is used (see ch.6).

A heat exchanger in the storage tank invariably means loss in efficiency. Only between 60 and 90% of the heat absorbed by the collectors will reach the water in the storage tank through the exchanger. There are two alternatives to the indirect anti-freeze solution:- insulating covers over the collectors on cold nights and draining the collectors. Both have their own disadvantages. Covers have the attraction of simplicity and economy. They also require a good memory and self-discipline. To

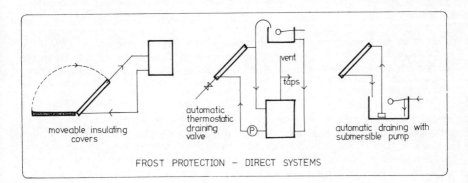

moveable insulating covers

automatic thermostatic draining valve

vent

taps

automatic draining with submersible pump

FROST PROTECTION — DIRECT SYSTEMS

lay in bed on a sunny morning will (more than ever) be a waste of useful energy, whilst to forget to cover on a cold night could be a costly mistake. Harold Hay, an American who is famous for building rooftop ponds to temper room temperatures in climates which are very hot by day, and very cold by night, uses removable insulation panels over his roof ponds. He claims that people will learn to use them just as they learn to put on their coats in cold weather. His roof pond systems, however are not liable to be damaged if they were allowed to freeze. In any case removable covers can only be a practical solution when the collectors are in an accessible position.

Draining the system can be carried out in two ways. The simplest is to use a submersible pump and arrange the storage tank so that the water from the collector flows back into the tank by gravity when the pump is not operative. Otherwise a thermostatic valve can be fixed to the lower end of the collector and set to open when the temperature in the collector drops below five degrees centigrade.

Direct systems, in which the water coming out the hot taps has actually passed through the collectors, are more liable to corrosion. Rusting for example occurs most readily where iron or steel comes in contact with water and air. There is always some air dissolved in tap water. In indirect systems, however, the amount of air present in the system is limited and once used up in initial surface rusting, it will not be replaced until the system is drained. In a direct system there is a continual flow of fresh water into the collectors and hence a continual supply of corrosive solution and mineral deposits. Direct systems should only be contemplated in installations where the collector is made of a highly corrosion-resistant material such as copper or stainless steel. Central heating radiators made of pressed steel or aluminium will develop pinhole leaks within a few years and sometimes much less if connected in a direct system. In areas where there is a high concentration of carbonaceous compounds, as indicated by the presence of furring in kettles, an indirect system is advisable to avoid blockages in the pipe work.

If there is an existing hot water system in the house, one could simply add a second hot water tap for the solar heated water. A more sophisticated solution would be to integrate the two systems with the solar system acting as a pre-heater and the conventional system operating only when the solar heated water is not as hot as required.

This might be done with instantaneous heaters positioned at each outlet. When the solar heated water is hot enough to use, the heater could be turned off. When the temperature of slightly heated water needs boosting there are complications however.

Instantaneous gas heaters have controlling mechanisms which regulate the quantity of gas to be burned in relation to the volume of water flowing through the heater and not in relation to the temperature of the water. What is needed is a valve on the gas feed pipe controlled by a temperature sensor on the water feed pipe. It would be wise to seek the advice of the manufacturers before attempting such an alteration. Simple adjustment can be made by varying the rate at which you run the tap but this is a rather hit-and-miss method.

Electrical instantaneous heaters often have a knob which allows one to vary the output temperature. This would be handy if it were to be fed pre-heated water. It is a simple matter to integrate the solar system with an instantaneous heater. The cold water supply is diverted from the unit and led into the bottom of the solar storage tank. The feed for the heating unit is then taken from the top of the tank. Warm water tends to rise. The top of the storage tank is therefore always the best position from which to draw off.

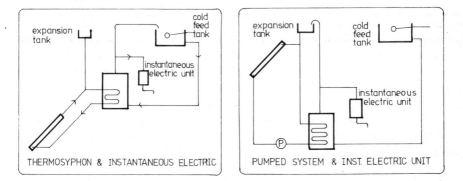

THERMOSYPHON & INSTANTANEOUS ELECTRIC

PUMPED SYSTEM & INST. ELECTRIC UNIT

The attraction of instantaneous heaters is that they come into operation only when the hot tap is opened. There is no wastage therefore due to heated water being left unused. However, the difficulty in automatically adjusting their fuel consumption when operating with a pre-heated supply which varies in temperature means that storage heating systems ought to be considered.

The storage tank systems are heated by an electrical immersion element or by a separate boiler circulating hot water through the tank either directly or indirectly via a heat exchanger. The heat input to the tank is controlled by a manually operated switch or automatically by a thermostat or time switch. Solar pre-heating can take place in the existing hot tank or in a separate solar storage tank, which then feeds into the existing tank.

If the solar collectors are connected to the existing storage tank, there will obviously be savings in both spending and space. This system will work best with tall, vertical tanks in which a clearly defined thermal gradient will occur — hot at the top and gradually cooler towards the bottom. The conventional heating source must be placed in the upper half of the tank, above the solar input. This may mean the existing plumbing will need to be altered. Otherwise there is the danger that the whole tank will be heated in the normal way and the solar collectors will never become hotter than the tank temperatures. In such a case the solar system would never begin to circulate. If the conventional heater is raised to the upper half of the tank, this does of course mean that the hot water storage capacity, except in sunny

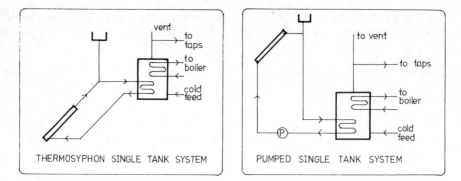

THERMOSYPHON SINGLE TANK SYSTEM

PUMPED SINGLE TANK SYSTEM

periods, will be effectively halved. The single tank system will only perform satisfactorily then if the existing tank is a large size, e.g. greater than 180 litres (40 gallons) in an average size family house. Chapter 6 includes a table showing tank dimensions and capacities and also illustrates the means of connecting new heat exchangers in old tanks.

More efficient solar collection, additional hot water storage capacity and simpler installatiion are possible if one adopts a separate solar storage tank. The cold feed from the existing hot tank is diverted to the bottom of the solar tank which then feeds from the top, back to the bottom of the first tank.

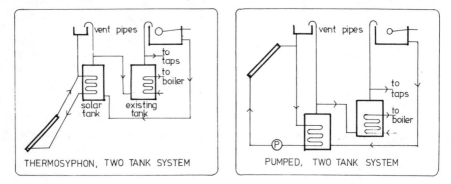

THERMOSYPHON, TWO TANK SYSTEM

PUMPED, TWO TANK SYSTEM

## Conclusion

This chapter has indicated the variety in system design. Other layouts which have not been shown may also be possible. The two tank system has been described as the most efficient. It is also expensive. It is important to study your existing water heating system before you decide on a solar system because obviously savings can be made by using what is already available to the best advantage. The next chapters describe plumbing in more detail and chapter 9 in particular might help you to understand better your existing heating system.

# 8

# Plumbing Techniques

The name "plumber" comes from the latin word meaning lead, and that's what the original Roman plumbers used for everything. Piping, drainage, connections, everything was made on site from sheet lead.

Thinking of the task facing the ancient plumbers always makes me very relieved to be living in an era where ingenious components bring simple plumbing work within the scope of just about anyone capable of wielding a spanner. Certainly the plumbing of a solar water heating system could be tackled by a home handyman or woman, given a knowledge of a few basic principles and techniques and an awareness of the components available.

This chapter describes the materials suitable for the job and illustrates the simplest, the conventional and the most economical techniques for connecting pipe to pipe, and pipe to tank. Readers with plumbing experience might proceed to the next chapters which indicate sizes and layouts, and examine the effect of different solar collector connection sequences.

## Materials

There are five kinds of pipe which might be used in solar water heating:
uPVC rigid plastic — for cold water only.
Polythene (alkathene) flexible plastic — for cold water and temporary outdoor heated water.
Nylon — hot water.
Copper — hot and cold.
Stainless Steel — hot and cold.

## UPVC and Polythene

The plastic pipes are recommended only for cold water. In practice polythene has often been used for connecting solar collectors to storage tanks. This means the pipe may sometimes be carrying water hotter than 90 deg. C. which would have the effect of softening the polythene and eventually one might expect the joints to leak under these conditions. For indoor pipework therefore, hot water connections in polythene are ill advised. Outdoors, where the occasional leak will cause no damage the use of polythene piping can lead to considerable savings on the initial outlay and ease of installation. Polythene piping (BS 1972 class C) is a flexible thick walled black pipe which should not be confused with plastic garden hose pipe which will only last for a few months when subjected to continual mains pressure.

uPVC is cheaper than copper for cold water piping particularly where there are long straight runs of piping. If there are a large number of bends polythene class C 1972 which is flexible, might work out cheaper still.

## Nylon

Nylon piping can be used for hot water. It is usually available only in small diameters — 15 mm. and smaller from DIY central heating companies such as Amkit. The small diameter means that its uses would be restricted to linking the solar collectors to the storage tank in pumped systems. The pipe itself is cheaper than copper but the proprietary fittings for it are dearer than those for copper.

## Copper

Copper piping can be used for any purpose. Being easy to work with and least prone to corrosion, it is the most popular choice.

## Bending Copper Pipe

Copper can be bought in diameters from 6mm upwards. When the pipe is smaller than 15mm. diameter, it can be easily flexed by hand to make large radius bends. A small tool is available for making tighter bends. 15mm. and 22mm. pipe can be bent across a formed block of wood or even around the knee if care is taken. A bending spring is necessary for this operation to distribute the bending force and prevent cracks forming. The spring must fit snugly inside the pipe at the point where the bend is required. A different spring is therefore required for each diameter of pipe used.

A cord should be tied to the end of the spring to enable withdrawal after the bending is completed. If it becomes stuck, it can usually be dislodged by slightly bending back the pipe. Tapping the pipe and where possible, twisting the spring will also help to loosen it.

Bending should always be a gradual movement. Jerks will result in wrinkles or even cracks. If difficulty is encountered in bending 22mm. pipe by hand, the work can be eased by heating the pipe.

Pipes of diameter larger than 22mm. require the use of a bending machine if the pipe is to be formed. This is an expensive piece of equipment and the only situation in which pipes larger than 22mm. would be required is the connection between solar collectors and tank in a gravity circulation system. Where a bending machine is not available, right angle and oblique bend fittings could be soldered or compressed on to the pipe.

## Stainless Steel

Stainless steel can be used in any situation where copper is used. It is harder than copper though and sizes greater than 15mm. (½") cannot be bent by hand. It is not so widely available as

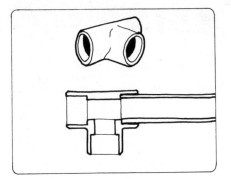

8.1 uPVC tee fitting.

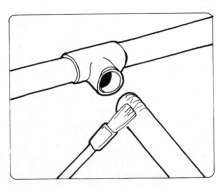

8.2 Solvent welding uPVC pipe.

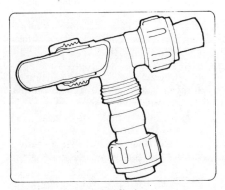

8.3 Compression tee fitting.

copper but is usually slightly cheaper.

### Jointing

The fittings for uPVC are solvent welded, i.e. they are simply glued to the pipe. For copper, stainless steel and polythene pipe, compression fittings can be used. These are expensive but allow one to carry out all the plumbing work without any soldering or use of a blowtorch. Copper and stainless steel can also be joined by means of capilliary ("Yorkshire") fittings and end feed fittings. These joints are made watertight by melting solder and allowing this to form a bond between the pipe and fitting. This operation requires the use of a blowtorch.

### A Spanner for the Works

Plumbing is usually associated with blowtorches and soldering. Nowadays, using compression fittings it is possible to complete a plumbing system using only a pair of adjustable spanners, a hacksaw, and a file.

Compression fittings consist of a brass body, brass nuts and soft copper rings, known as olives. These rings are passed round the pipe to sit between the pipe and the fitting body. Above the olive is a nut, also passed round the pipe. When this nut is tightened onto an external thread on the fitting body, it compresses the olive, tightening it round the pipe and forming a watertight joint.

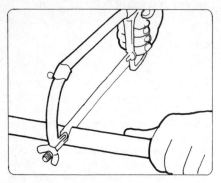

8.4 Cutting pipe with hacksaw.

1) Cut the pipe at right angles using a hacksaw or pipecutter.

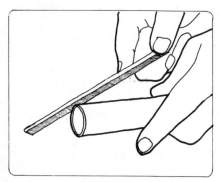

8.5 Removing rough edges with a file.

2) File off any burr left on the end of the pipe.
3) Check that the end of the pipe is circular in section and not deformed by dents.
4) If the pipe is deformed, you will have to begin again cutting off the non-circular section of pipe.
5) Use the file to scratch a small mark 22mm. (¾") from the end of the pipe. This will serve as an indicator of how deep the pipe is inserted in the fitting.

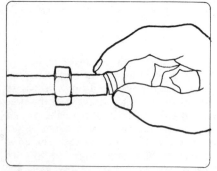

8.6 Compression nut and olive are fitted over pipe end.

6) Loosen the nut on the fitting by turning it anti-clockwise and remove it together with the compression ring or olive which sits between the nut and the fitting.
7) First the nut and then the olive should be slid over the end of the pipe.
8) Push the pipe into the fitting till it comes to rest on a shoulder about 15mm. (½") inside the fitting.

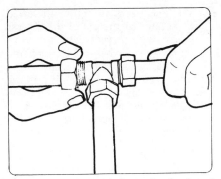

8.7 Pipe is inserted into the body of the compression fitting and the nut is tightened by hand.

9) Hand-tighten the nut over the thread on the fitting. This will compress the olive between the pipe and the fitting.

10) The joint is made watertight by tightening the nut with a spanner by the amount shown—
15 mm pipe — 1¼ turns
22 mm pipe — 1 turn
28 mm pipe — ¾ turn

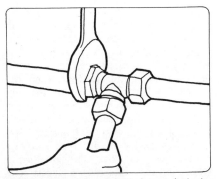

8.8 Compression joint is sealed by tightening nut with a spanner.

Notes:
a) The joint can be disassembled simply by loosening off the nut.
b) Once tightened with a spanner, however, the olive will be permanently tightened around the pipe.
c) There are several makes of compression fittings and it is worth remembering that the nuts from one manufacturer will not necessarily fit the body of another manufacturer's fittings.

**Joining Polythene Pipe with Compression Fittings**

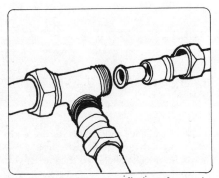

8.9 Compression tee fitting for polythene pipe.

Polythene pipe is most easily cut with a hacksaw. Thereafter the process is the same as that described for copper pipe with the addition of one step.

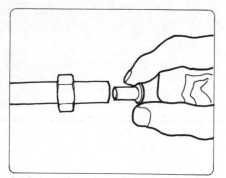

8.10 A copper sleeve is pushed into the mouth of the polythene pipe in addition to fitting the nut and olive.

7A) A copper sleeve must be inserted in the end of the polythene. This is necessary to prevent deformation of the polythene when the compression nut is tightened. Offcuts from copper tubing can be used for this purpose but it is advisable to use purpose-made sleeves which have collars to prevent them slipping too far into the polythene pipe.

## Plumbing with a Blowtorch

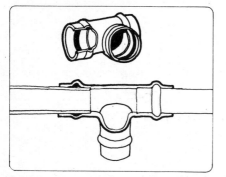

8.11 Capilliary tee fittings have a ring of solder in each opening.

Capilliary ("Yorkshire") fittings are considerably cheaper than compression fittings and it requires little skill to make a successful job with them. The job is made easy by the fact that a solid ring of solder is integrated into each socket of the fitting. This means that the fitting and pipe simply have to be cleaned and smeared with plumber's flux, joined together and heated. The flux is necessary in order to help the solder flow along the joint and form a bond between the two surfaces. The joint can be heated above a gas cooker, but a blow torch will be much more convenient. The modern blow torches which use small butane gas containers are much easier to work with than the old petrol ones.

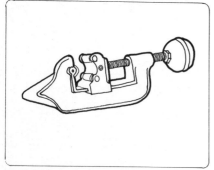

8.12 Pipecutter.

## Joining Copper Pipe with a Capilliary Fitting

1) Cut the pipe with a hacksaw or pipecutter. The cut should be made at right angles to the run of the pipe.
2) If a pipecutter is used, the pipe end will have been turned in and

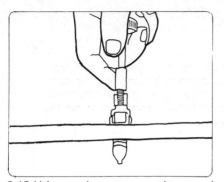

8.13 Using a pipecutter: revolve round pipe, whilst slowly increasing blade pressure by twisting handle-knob.

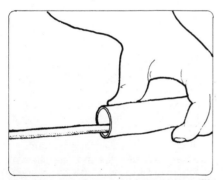

8.14 Removing burr with round profile file.

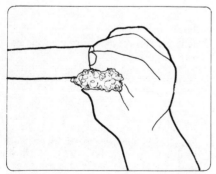

8.15 Brushing pipe end clean with wirewool.

should be opened again by inserting the blunt end of the cutter and twisting it. Care should be exercised so as not to flare out the end of the pipe such that it might not be easily inserted into the fitting.

3) File any burr off the end of the pipe.

4) Brush the end of the pipe clean with wire wool so that it shines.

5) Similarly brush clean the inside of the fitting.

6) Smear the pipe and the inside of the fitting with flux.

7) Insert the pipe into the fitting and push it until it comes to rest on a shoulder about 15mm. (½") inside.

8) All other branches of the fitting should be similarly prepared before applying heat.

9) Using a blowtorch, gradually heat the whole fitting. First the flux will be seen to bubble out from the joint and after a few minutes a silver ring of solder will appear around the pipe at the mouth of the fitting.

10) When a complete silver ring is visible around each branch of the fitting, remove the flame and leave the fitting undisturbed for a few minutes to allow the solder to solidify.

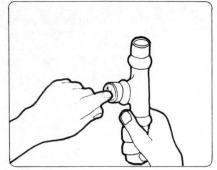

*8.16 Smearing flux on the contact surfaces after cleaning with wirewool.*

a)  The joint can be disassembled simply by reheating until the solder is molten and then pulling the pipes apart.

b)  Once disassembled a fitting can be reused in the manner described for end-feed fittings.

c)  If it is necessary to solder one branch of a fitting before the other branches are prepared, a length of copper pipe, uncleaned and without flux, should be inserted in the branch not yet prepared. A wet rag must then be wrapped around this pipe in order to absorb the heat and prevent the solder in that branch from melting.

*8.17 Heating the assembled joint with a blowtorch.*

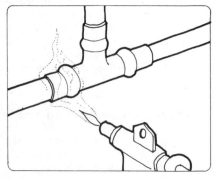

d)  Bending a pipe after it is soldered can weaken the joint. Any bending should therefore be done before soldering.

e)  If the joint leaks when the pipe is filled with water, drain it down and reheat the joint. Extra solder can be added round the mouth of the fitting.

## Joining Copper Pipe with End Feed Fittings

These fittings are similar to capilliary ("Yorkshire") fittings lacking only the silver ring of solder inside. The procedure is therefore the same as with capilliary fittings till step 7.

7) "Tin" pipe end with solder.

8) Insert pipe in fitting.

9) Holding a blowtorch in one hand gradually heat the fitting. In the other hand hold a roll of plumbers' solder wire.

10) After a few minutes, when the copper begins to change colour with the heat, dab the solder wire first in a tin of flux, then touch it on the mouth of the fitting. If the copper is hot enough, the solder will melt instantly and capilliary attraction will draw it into the crack formed between the pipe and the fitting.

11) Continue adding solder round the whole circumference of the pipe till a complete circle of solder is formed around the mouth of the fitting. The joint should then be complete.

## Joining Pipes with Threaded Joints

These are used mostly where pipes connect with a tank. The threads are distinguished as being either external (male) or internal (female) threads. The watertightness of the joint depends on the prevention of water seeping through its threads. This is achieved by the use of a sealing paste such as Boss white or P.T.F.E. tape which is wound round the external thread.

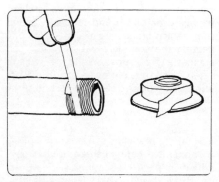

*8.18 PTFE tape is wound round the external thread to seal the joint.*

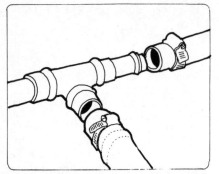

*8.19 Hose-clip joints between copper and polythene tubes.*

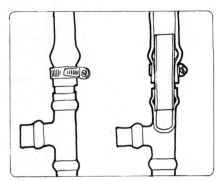

*8.20 External and cross-sectional view of copper/polythene joint showing soldered olive strengthening the grip.*

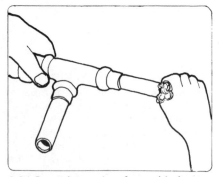

*8.21 Preparing tube for soldering on olive: brushing clean with wirewool.*

## Backyard Plumbing Techniques

In addition to the methods already described, there are various materials and techniques which can be used for experimentation, temporary or backyard installations where economy is the most pressing factor and an occasional leak will not cause much damage. Cheap plastic piping can be used for long runs and short sections of old rubber hose from automobile engines for connecting absorber panels.

It is best to avoid using too many different diameters of piping and to have the minimum number of joints as it is always these points which fail first.

The most successful joint between two plastic pipes is made by sliding them over opposite ends of a short length of tight fitting copper pipe and making the joint fast by tightening a jubilee clip (hose clip) over the plastic. If there is much pressure on the joint, it will be necessary to first solder an olive onto the ends of the copper pipe. This will provide a grip for the plastic pipe. The jubilee clip should then be fixed on the end of the plastic pipe which has been slid over the olive. On pipes which do not carry drinking water, a further precaution would be to add a sealant between the plastic and copper. Various proprietary brands are available for connecting the cooling pipes in motor car engines. (N.B. Boss White, a sealant often used by plumbers on threaded joints is incompatible with certain plastics).

For taking branches off a run of piping, a copper tee fitting can be used as described in previous sections. This would be made up with short lengths of copper tube each having an olive soldered

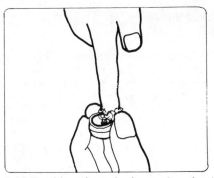

8.22 Brushing clean the internal surface of olive prior to soldering.

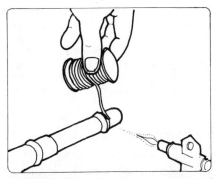

8.23 Soldering olive on to the end of copper tube.

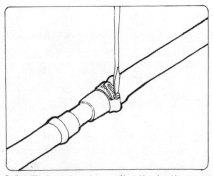

8.24 Tightening hose (jubilee) clip over polythene tube which has been slid over the olive on copper tube.

on the end and then the plastic pipes could be joined as described above.

It is worth remembering that plastic piping will be degraded by the action of sunlight. It is therefore advisable to have the piping shaded from the sun. Carbon black is sometimes added to resist the effect of sunlight. Hence black plastic is normally the best colour choice.

## Joining Plastic Pipe with Jubilee Clips

1) Select an offcut of copper pipe at least 150 mm. (6″) long and free from dents at the ends. (If the pipe section is not circular it will be more difficult to make a watertight joint). In order to provide a tight fit, a copper tube with the same internal diameter as the plastic should be selected.

2) Clean the ends of the copper with wire wool and also the internal surfaces of two copper rings (olives) which fit over the ends of the pipe.

3) Smear the tube and olives with flux.

4) Gripping the pipe in a vice or long handled pliars, heat the ends.

5) After a few minutes, when the pipe is hot, touch it with the end of a roll of solder wire dipped first in flux.

6) If the pipe is hot enough to melt the solder on contact, continue to apply solder in a circle about 22 mm. (¾″) from the end of the pipe.

7) Wipe the solder smooth with a plumbers' cloth or stout brown paper. (This is called "tinning").

8) Do this at both ends.

9) Slide the olive onto the pipe and locate it over the soldered circle.

10) Apply heat to the olive.

11) Feed solder into the crack between the olive and the pipe.

12) Similarly solder an olive on the opposite end.

13) Leave the copper undisturbed to cool.

14) Position two jubilee clips in the centre of the copper tube.

15) Dip the plastic pipes into boiling water in order to soften the ends.

16) Force them over the ends of the copper tube till they overlap the soldered olives by at least 25 mm. (1").

17) Slide the jubilee clips over the end of the plastic pipes and tighten them with a screwdriver.

## Plumbing Principles

There are three principles involved in plumbing solar systems which if properly appreciated will help you work out all your plumbing requirements. These are:

(i) Water finds its own level.

(ii) Air is dissolved in water and it comes out of solution when heated and floats up to the highest position it can reach.

(iii) Warm water tends to rise above cold water in convection currents.

## Water Levels

If you connect a pipe to a tank full of water, and take that pipe round an obstacle course of bends, vessels and other tanks, the water flowing into the pipe will always try to reach the same level as the surface of water in that first tank. If the end of the pipe is held below this level, there will always be pressure there, unless the tank is allowed to empty itself through the pipe and thereby equalise its level on the ground. Only when the pipe end is lifted above the tank water level will the pressure cease and hence water would stop flowing out through it.

## Dissolved Air

To continue with the example of a pipe connected to a tank of water, if the pipe end was held below the level of water in the tank and water did not flow out through it and there were no kinks in the pipe, we might be almost certain that there were air blockages in the pipe. Air is present in pipework before it is filled and can also appear after filling if the water in the pipe is heated. This happens in the same way that air bubbles appear in a saucepan of water being boiled. Because the air bubbles are less dense than water, they attempt to float up to the water surface. In plumbing systems, however, they are often trapped by bends in the pipework. Such points where air is trapped must be fitted with air vents, either automatic or manual, so that the air does not form a blockage to the water circulation.

## Convection

The tendency of warm water to rise above cold water has been mentioned several times and is relevant to the working of gravity circulation systems and the locating of hot water draw-offs.

## Conclusion

This chapter has described the techniques and principles you need to know to tackle your own plumbing. The next chapter will detail the procedures involved in putting your new knowledge into practice in the installation of a solar water heating system.

# 9

# Plumbing Installation

This chapter shows how you can put to use the principles and techniques which were explained in the last chapter. It is always easiest of course to learn by example and, fortunately, we are all surrounded by examples of applied plumbing. Let's begin by taking a look at the existing hot and cold water supplies in your own home.

Nowadays, cold water is delivered under pressure from an underground mains service pipe to most homes. Some cold water outlets, such as the kitchen tap, may be taken directly from the mains pipe, but most draw-off points will be taken via a large cold water storage tank in the attic or in an upper floor cupboard. In purpose-built blocks of flats, there is often a shared tank on the roof-top. The tank is filled through a ball-cock. This is a valve controlled by a plastic ball float on the end of a lever arm, just like the device inside a W.C. cistern. The valve closes whenever the level of water in the tank, and hence the ball float, reaches a predetermined height, and re-opens when water is drawn-off. There should be an overflow pipe to safely dispose of water should the ball-cock ever fail to cut off the mains supply.

If you are not already familiar with your hot water system, it is best to be very methodical. Starting at the hot tap, work your way around the house following the pipework. When you come to tanks, look, or feel all round them so that you do not miss any connections. Now make a diagram of the system. You do not need to draw all the bends, or even get the scale right. But you should be able to show exactly what is connected to what, and the relative heights of fittings and the relative positions of tank connections.

Different houses have different plumbing systems but the most common layout is that shown in figure 9.1. A cold storage tank feeds a copper cylinder below it. Inside the cylinder, water is heated by an electrical immersion element or by a heat exchanger through which hot water from a boiler is circulated. Note the vent pipes taken from the highest points of both the boiler circuit (primary) and the hot water distribution circuit (secondary). These are necessary to allow for the escape of air bubbles and the expansion which occurs when water is heated.

A solar pre-heating system can be integrated with a hot water system by adding a solar tank between the cold storage tank and the hot water cylinder. The

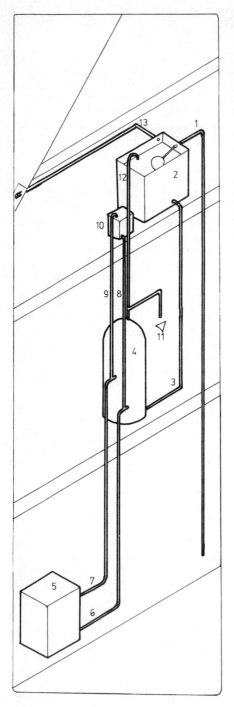

solar tank is heated by solar collectors in the same way that the existing cylinder is heated by the boiler. On very sunny days the boiler will not be required at all. On less favourable occasions, the boiler will boost the temperature of the water leaving the solar tank before it is delivered to the taps.

Figures 9.2 and 9.3 show the layout of both a gravity-circulation thermosyphoning and pumped solar systems integrated with the standard hot water system. Try drawing a diagram of how a solar system could be linked in with your own hot water plumbing.

**Conventional Boiler Hot Water System**

1. *Mains supply*
2. *Cold storage tank*
3. *Cold feed to hot water cylinder*
4. *Hot water cylinder*
5. *Boiler*
6. *Return to boiler*
7. *Flow from boiler to hot water cylinder*
8. *Feed to boiler circuit*
9. *Vent and expansion for boiler circuit*
10. *Expansion tank*
11. *Hot water distribution to the taps*
12. *Vent and expansion for hot water cylinder*
13. *Overflow from cold water storage tank*

# THERMOSYPHONING (GRAVITY) INDIRECT, TWO-TANK SOLAR PRE-HEAT SYSTEM

1. Solar storage tank
2. Cold feed to solar tank (22 mm)
3. Cold feed tank (10-15 galls) (45-70 litres)
4. Overflow (22 mm)
5. Mains supply (15 mm)
6. Solar tank vent (22 mm)
7. Pre-heated solar draw-off (22 mm)
8. Primary flow from collectors (28 mm)
9. Primary return to collectors (28 mm)
10. Primary vent (22 mm)
11. Primary feed (15 mm)
12. Expansion tank (4 galls) (20 litres)
13. Solar collectors
14. Extended vent from existing hot water cylinder (22 mm)
15. Hot water to taps (22 mm)

**PUMPED, INDIRECT, TWO-TANK SOLAR PRE-HEAT SYSTEM**

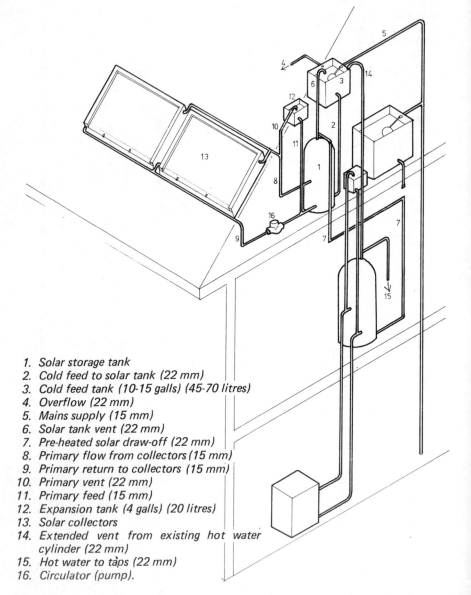

1. Solar storage tank
2. Cold feed to solar tank (22 mm)
3. Cold feed tank (10-15 galls) (45-70 litres)
4. Overflow (22 mm)
5. Mains supply (15 mm)
6. Solar tank vent (22 mm)
7. Pre-heated solar draw-off (22 mm)
8. Primary flow from collectors (15 mm)
9. Primary return to collectors (15 mm)
10. Primary vent (22 mm)
11. Primary feed (15 mm)
12. Expansion tank (4 galls) (20 litres)
13. Solar collectors
14. Extended vent from existing hot water cylinder (22 mm)
15. Hot water to taps (22 mm)
16. Circulator (pump).

The plumbing work for a solar collector centres around the solar storage tank. This has to be connected to:

a) a cold water supply
b) the hot water draw off points (secondary circuit)
c) the solar collectors (primary circuit).

## COLD WATER FEED

If there is an existing cold water storage tank at least a foot above the level of the top of the solar storage tank, this can be used as the cold water feed tank. Otherwise a new tank must be installed. The cold feed tank should be connected to a mains water supply via a ball cock similar to the valve found in W.C. cisterns. An overflow pipe is also required in case the ball valve should ever fail to turn off the mains.

### Installation of Cold Feed Tank

1) If the tank is to be situated in the attic, measure the length and breadth of the access;
2) Select a plastic cold water tank whose actual capacity is equal to at least half the capacity of the hot water cylinder, ensuring that it will fit through the attic access;
3) Prepare a level platform for the tank. Plastic tanks must be fully supported on their base and cannot span between two bearers like galvanised tanks.
4) Select a mains pressure ball cock for a cold water storage tank. When buying this item say that it is for a plastic tank and that you need two nylon washers with the fittings.
5) Purchase a 22 mm. (¾") P.V.C. overflow fitting with two nylon washers.

*9.4 Cold feed tank, showing mains supply, cold feed pipe and vent pipe.*

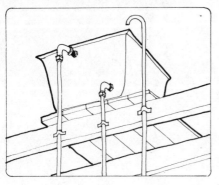

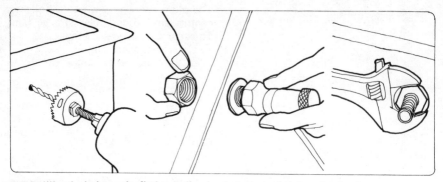

*9.5 Drilling hole in tank, fitting ball-valve in hole and tightening back-nut.*

6) Purchase a 22 mm. (¾'') or 28 mm. (1'') tank connector for plastic or copper pipe.

7) Using a tank cutter, drill a hole in the side of the tank some 75 mm. (3'') from the top of the tank to accept the threaded shank on the end of the ball valve. The tank should be supported from underneath when drilling. If you do not have a drill bit of the correct size a circle of smaller holes can be made and a round section file used to smooth down the ragged edge. It is important that you end up with a tight-fitting, smooth edged, circular hole.

8) Similarly prepare a hole to fit the overflow 50mm. (2'') beneath the top of the tank.

9) Similarly prepare a hole for the tank connector 60 mm (2½'') above the bottom of the tank. This will connect to the pipe carrying water to the solar cylinder.

10) Remove the nut from the shank of the ball valve.

11) Slide one of the nylon washers over the shank.

12) From the inside of the tank, insert the shank of the ball valve through the prepared hole.

13) From the outside of the tank slide a nylon washer over the shank.

14) Thread the nut over the shank and tighten over the washer.

15) Use a spanner to hold the fitting steady on the inside of the tank whilst using another spanner to

*9.6 PTFE tape is wound around the ball valve shank and a female iron to 15 mm compression elbow is fitted which will accept the rising main pipe.*

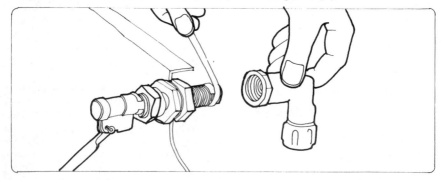

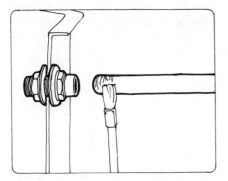

*9.7 Overflow connector is fitted to the tank and ready to accept the PVC overflow pipe after both contact surfaces are cleaned, sanded and smeared with solvent weld solution.*

*9.8 Cold feed to solar cylinder enters via a 22 mm elbow with a drain-cock, a 22 mm copper to male iron couple and ¾" to 1" bush.*

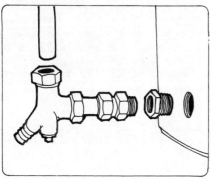

finally tighten the nut on the outside.

16) Wind PTFE tape, or smear boss white round the threaded shank and fit the pipe connection. This should be either a straight couple or a bend, depending on the location of the mains supply, with an internal thread at one end and a compression or capilliary socket to receive the pipe at the other end.

17) Similarly fit the overflow and tank connectors. If using compression fittings, it is advisable to remove the nut and olive before fixing the fitting body to the tank. Otherwise the olive may be accidentally compressed and hence become too small to fit over the pipe.

18) Locate the tank on the platform.

**Connecting an Overflow Pipe**

22 mm. (¾") P.V.C. piping is usually used for this purpose. The pipe is joined with solvent welded fittings which can be bought at plumbers merchants. It can discharge over a roof, gutter, drain or simply project from the wall of the house, preferably over porous ground. If there is an existing cold water tank at lower level than the cold feed tank, and this already fitted with an overflow pipe, the cold feed overflow could simply discharge into this tank.

**Connecting the Cold Feed to the Solar Tank**

This pipe should be of large enough diameter to allow the solar tank to refill as quickly as water is drawn off from it. Otherwise negative pressure will build up in the cylinder and water will come out of the hot taps in spurts. 22 mm. (¾") pipe is normally used, the larger

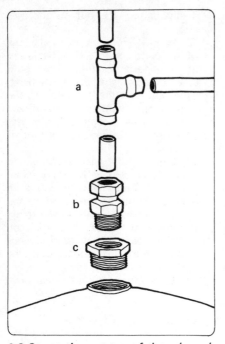

9.9 Connections at top of the solar cylinder. Expansion/vent pipe rises straight upwards and hot water distribution is taken from this via a tee. a) 22 mm capilliary tee; b) 22 mm compression to ¾" male iron couple; c) ¾" to 1" bush.

9.10 A foam insulation sleeve — remember to tape along the joints.

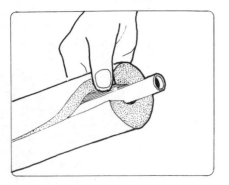

diameter being used when there is a long pipe run between feed tank and cylinder or when there is a likelihood of many hot taps being run simultaneously. The cold feed to the solar cylinder should always enter the lowest connection on the tank.

## HOT WATER DRAW-OFF

As water in the solar cylinder is heated, there are two physical reactions which must be allowed for. Water expands when heated and air dissolved in the water tends to come out of solution just as bubbles appear in a saucepan of boiling water. A vent pipe connected at the highest point on the cylinder, and open at the end, will allow the air bubbles to escape without causing blockages. The vent pipe will also, depending upon its length and height, take up the expansion which occurs in the water. The pipe should, however, be bent over the top of the cold feed tank so that any extra expansion will overflow safely into the tank. A 22 mm. (¾") copper pipe is required for this purpose.

The hottest water rises to the highest point in the cylinder. The draw-offs are usually taken from a tee-piece inserted in the vent pipe just above the cylinder. If, as indicated in Ch. 7, the solar heated water forms a pre-heated feed for the conventional hot water cylinder, a draw-off pipe of equal diameter to the cold feed pipe should be used (i.e. 22 mm. or 28 mm.). If the water is being fed directly to hot taps, 22 mm. pipe will suffice.

Considerable heat loss occurs from hot water distribution pipes and hence they should be kept as short as possible and, of course, should be well lagged.

If there is an existing cold feed pipe in the attic, connecting a cold water storage tank to an existing hot water cylinder, the hot draw off from the solar cylinder can economically be connected to this pipe. There should be a stop cock on the existing pipe. A tee-fitting should be inserted on the downward side of this to accept the solar preheated water. A stop cock on the solar heated branch will mean that, should future circumstances require it, the solar system can be isolated and the conventional cylinder fed by the cold tank as before.

1) Measure diameter of existing cold feed pipe and purchase an appropriate tee-piece.

2) Close the stop cock on the cold feed pipe from storage tank to hot water cylinder.

3) Mark a location for a tee to accept solar preheated water into cold feed between the stopcock and the hot water cylinder.

4) Drain off the feed pipe to a level below the marked section by means of a drain off cock which should be located at the lowest point on the pipe. If the top of the hot water cylinder is below the section where the tee will be inserted, the pipe can be sufficiently drained simply by runnine the hot taps till they run dry.

5) Using a hacksaw cut out a section about 30 mm. long. Long enough to accept the tee piece.

6) Connect the tea piece and the solar preheated branch in the manner described in plumbing techniques.

7) Close the domestic hot water taps.

8) Open the stop cock on the solar preheated branch, leaving the cold feed stop cock closed.

*9.11 Connecting solar cylinder to conventional cylinder via existing cold feed pipe.*

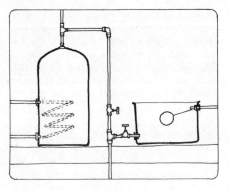

If you have installed a new cold feed tank at a higher position than the former cold storage tank over which the vent pipe from the top of the existing cylinder discharges, it will be necessary to extend this vent to discharge over the new cold feed tank. Otherwise the force of gravity will cause the vent to continually overflow.

## CONNECTING THE SOLAR COLLECTORS

A flow pipe is required to connect the highest point of the solar collector to an inlet on the solar cylinder which should be about half way or two thirds up the side of the cylinder. A return pipe is also required to carry cold water from the cylinder to the bottom of the solar collector. The size of these pipes should depend on the system of circulation, the length of the pipes, number of bends in the pipes, and type of pump, if any, and the size of the solar collector.

With pumped systems 15 mm. pipe is usually used in connection with a normal central heating circulator (pump).

If the system is relying on gravity circulation (thermosyphoning), larger diameter piping is required. 22 mm. has been used successfully on small rigs but 28 mm. piping would allow the system to operate more efficiently as it would present less resistance to flow.

If the system is direct, i.e. the water flowing through the collectors passes directly through to the hot taps, then the expansion pipe on the solar cylinder

*9.12 Solar cylinder connections to a pumped, primary, indirect circuit (i.e. link between tank's heat exchanger and solar collector connecting pipes).*
*Note the location of the pump forcing water up the return pipe from the bottom of the cylinder. Also note the drain-cock at lowest point in primary circuit, and the cold feed pipe tee'd into the return pipe.*

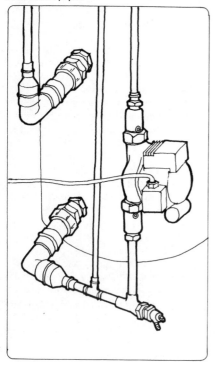

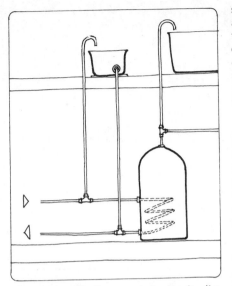

should accommodate water expansion and exhaust air bubbles. If however, there is a heat exchanger in the solar cylinder, a separate vent pipe will be required at the highest point on the flow pipe. In pumped systems the highest point will probably occur at the top of the collector.

In gravity (thermosyphoning) systems, the highest point should be where the flow connects to the solar cylinder. It is essential that the vent is connected to the highest point, otherwise air blockage will prevent heat being passed from the collector to the cylinder.

*9.13 In an indirect solar circuit (i.e where a heat exchanger is used) a vent pipe must be fitted to the flow pipe connecting collectors to cylinder, at its highest point. This discharges over an expansion tank which is connected by a feed pipe to the return pipe which delivers cool water to the bottom of the collectors.*

This vent pipe (22 mm. dia) should be positioned to discharge over a small expansion tank (4 galls/20 litres). This tank is connected to the return pipe thus doubling as a filler tank for the solar primary circuit. If the primary circuit contains anti-freeze, it is essential that the vent pipe discharges over this and not any other tank. Most anti-freeze solutions are toxic and must not be allowed to contaminate any of the other water tanks. This tank will be situated above the collectors and hence a platform for it is required at the top of the attic.

*9.14 Pump, connected to mains electricity via an electronic temperature differential control box.*

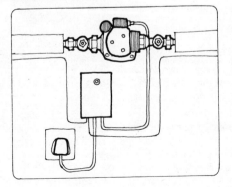

### Pump Installation

Before installing the pump, flush out the pipework with mains pressure to remove any loose particles which might damage its moving parts. It should be fitted between two isolating valves to facilitate future replacement. It must be wired to its control box which is in turn connected to the mains via a plug and socket or by a permanent connection to a switched and fused spur box.

# 10

# Collector Connection Patterns

The pattern in which collectors are plumbed together has considerable effect upon their performance. Ideally, we want water to flow through the different collectors, and through the different waterways in each collector, at an even velocity. If the flow through one particular area is comparatively slow, a hot spot will appear because the heat is not being transported away so rapidly as in the surroundings. Hot spots cause greater heat loss which results in lower operating efficiency.

This chapter explains the factors which affect the flow of water through a solar collector and provides a guide to selecting the best connection pattern for different types of system.

### Circulation Choice

The two different modes of circulation, gravity (thermo-syphoning) and pumped (forced), must be considered separately although flow through both types of system is affected by basic factors such as frictional resistance which is increased by long pipe runs and small pipe diameters.

### Gravity Circulation (Thermo-Syphon)

As has been explained in chapter 5, gravity systems rely upon the naturally occurring convective currents to lift the solar heated water upwards from the bottom of the collector to the top and from there to a storage at a high level. The convective force is not so very strong and hence all the plumbing connections must be made so as to facilitate the upward flow, and they should never be such as would force the heated water downwards in opposition to natural current. The outlet of each collector should be its highest point and header pipes are often tilted slightly away from true horizontal in order to enable this.

Thermosyphoning systems have lower flow rates and hence higher operating temperatures and the result is lower collection efficiency. They do however have the advantage of automatically self adjusting their flow rates to match immediate conditions. Local hot spots therefore are not so common; if the temperature does begin to rise in a particular area of the collector, this temperature rise itself will cause the flow rate through that area to increase thus limiting local heat build-ups. This self-regulating effect greatly simplifies the design of domestic thermosyphoning systems.

**Pumped Systems**

Pumped systems can achieve higher energy collection efficiencies, but more care is required in laying out the circulation route in order to realise the potential improvement. Attention should be paid to maintaining even resistance on all through-routes by matching pipe lengths and diameters. In addition there is a more subtle effect which comes into play in pumped systems. Varying pressure differences can build up across a grid of riser tubes between two header pipes which can minimise flow through certain waterways, or even reverse the expected direction of flow. These effects, which are not generally appreciated, were first brought to my attention by Kerr MacGregor of Napier College in Edinburgh. He had been working with a tube-and-sheet collector with a grid of pipes like that shown in figure 10.1. It had two 22 mm headers and six 15 mm risers between them. If you measure the length of each riser and the distance between its ends and the inlet and outlet, you will find the total length to be the same for each of the six possible routes through the collector. Thus the frictional resistance to flow will also be equal and you might therefore expect that the flow velocity through each riser would be the same (as indicated in figure 10.1).

In a thermo-syphoning set-up, this would indeed be the case. When water is pumped through the grid though, there is no flow at all in one or two of the central risers. Flow is most rapid through the riser nearest to the outlet (extreme right hand) and is actually in a downward direction in the riser nearest the inlet (Fig. 10.1).

*10.1 Expected flow pattern in a grid of pipes.*     *10.2 Actual flow pattern in a grid of pipes with pumped circulation.*

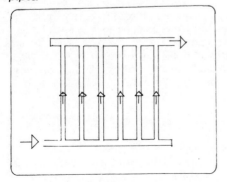

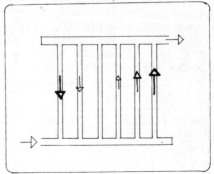

These unexpected results are due to the varying pressures which occur at the top and bottom of each riser. This pressure variation is in turn due to the difference in velocity along the length of the header pipe. On the bottom header velocity is at a maximum at the inlet and drops towards the right. In the top header velocity is at a maximum near the outlet, dropping towards the left. The greater the velocity of the stream across the entrance to the riser, the greater the tendency to draw water out of it. This is because the rapidly moving stream of water lowers the pressure at the riser entry (You may have heard this phenomenon described before as the Bernouli Effect).

In pumped systems therefore we have the situation that in the risers nearest the inlet, the pressure is higher at the top of the riser than at the bottom, hence water does not rise through the pipe, but flows downward instead. In the middle risers pressures at top and bottom are roughly equal and water actually rises at the desired velocity only in those risers nearest the outlet.

Mr MacGregor has suggested that a more even flow can be attained by having the inlet the outlet on the same side of the collector as shown in figure 10.3. In this

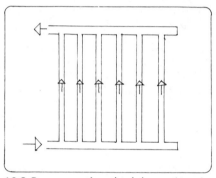

*10.3 By connecting the inlet and outlet at the same side of the absorber grid, in a pumped circuit, a more even flow may result.*

case the six different routes through the collector are all of different lengths. The variation this causes in frictional resistance to flow is small however compared with the circulation force supplied by the pump and hence flow is fairly even. At any rate, this is true in most collectors which have 15 mm risers. If the risers were smaller, say 8 mm, the first layout (figure 10.1) might apply as the increased frictional resistance could dominate the effect of varying pressure differentials.

### Series or Parallel

Having decided whether or not you will be using a pump for circulation, and having read above the requirements that your choice entails, you must then choose between connecting the solar collectors in series or in parallel.

## Connecting in Series

The longer the route which water takes in its passage through the collectors, the higher the temperature which it will reach. In some applications this may be desirable and can be encouraged by connecting a group of collectors in "series", i.e. connecting the outlet from the first collector to the inlet of the second and connecting its outlet to the inlet of the third etc. The final collectors in the series might be especially constructed for the higher temperatures at which they will operate e.g. with double glazing and additional insulation.

## Connecting in Parallel

The alternative to connecting in series, is in "parallel". In this case water is delivered to each collector by a header pipe which runs along their lower edge. The header can be external to the collector or an integral part of it. Heated water is withdrawn via a header along the upper edge of the collectors. Water therefore passes through only one collector and thus the operating temperature of the collector is lower. A lower operating temperature means that less of the collected heat will be lost thus the total energy collected by an array connected in parallel will be greater than that of a comparable array connected in series.

Note however that if a pumped system is connected in parallel between two headers, the "MacGregor effect" of varying pressure differentials which we have just discussed, again comes into play. In a parallel array of 4 collectors with inlet and outlet at diagonally opposite corners, (as in figure 10.4) the flow through the first collector might be in reverse whilst the second might have no flow through it at all. This would effectively reduce the total collection area. If the inlet and outlet were connected to the centre of the lower and upper headers (respectively), the flow would once again assume the more even flow distribution as indicated in figure 10.5.

*10.4 With diagonally opposite inlet and outlet on a parallel array, circulation through the collectors is likely to be uneven when pumped.*

*10.5 With a pumped, parallel array of solar collectors, a more even flow pattern can be obtained by locating inlet and outlet pipes centrally.*

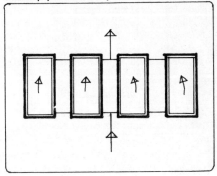

81

**Operating Temperature**

In situations where an auxiliary heating source is available to boost the water temperatures on days where there is insufficient sun, it is best to operate the collectors at a low temperature with the outlet water being only about 5°C higher than the inlet.

If there is enough radiation the system will continue circulating after the whole tank has been raised through 5°C. This cold water at 15°C must make several passes through the collectors to reach 45°–55°C, the temperature at which domestic hot water is usually delivered.

In installations where the sun is the only energy source it may be worthwhile operating at higher temperatures. A small quantity of really hot water is often more useful than masses of tepid water. Thus an array of collectors connected in series and linked to a tall, vertical storage tank in which temperature stratification is encouraged, might be the best choice for remote farmhouse without mains services, whilst an array of collectors linked in parallel would be more appropriate for a town house.

**Examples**

Different systems therefore require different solutions. Similarly, the collector construction will have great effect upon the internal flow pattern and hence upon the choice of connection layout. It is difficult therefore to make generalised recommendations, but the following diagrams will at least indicate the pros and cons of various alternatives.

**Collectors in Thermo-Syphoning Systems**

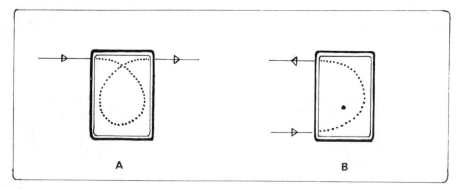

A                                    B

"B" is better than "A" because it introduces unheated water at the bottom of the collector. Introducing cold water at the top disturbs the convective currents.

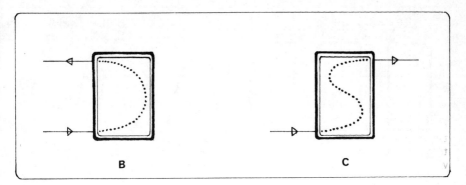

**B**  **C**

"C" is better than "B" because it will discourage the formation of 'dead' corners where no flow occurs.

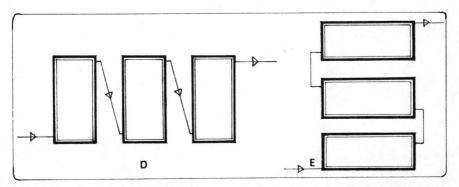

**D**  **E**

"E" is better than "D" because it allows the circulating fluid to rise continuously without forcing the flow downwards as in "D".

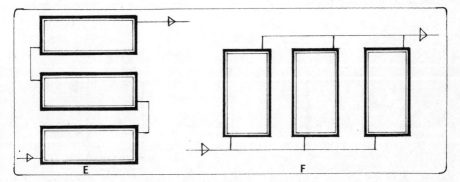

**E**  **F**

In some cases "F" is better than "E" because by allowing the water to flow 'in parallel' between two header pipes, it will operate at somewhat lower temperatures and hence at higher efficiencies. If high temperatures are required however the 'series' connection in "E" will be preferable.

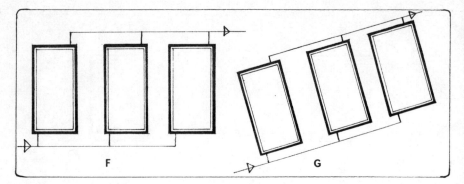

"G" is better than "F" because by giving the horizontal headers a slight tilt towards the outlet, it reduces the resistance to the convective force. The water will therefore circulate faster and more energy will be collected.

## Collectors in Pumped Systems

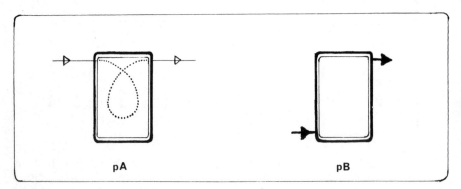

"pB" is best because there is less likelihood of dead areas of no-flow occuring.

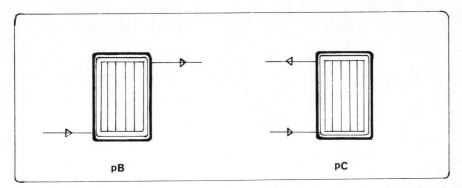

pB            pC

With header-and-riser type absorbers, *"pC"* is better than *"pB"* because varying pressures build up in the headers of *"pB"* and disturb the flow through it. This is the 'MacGregor effect' described in the text.

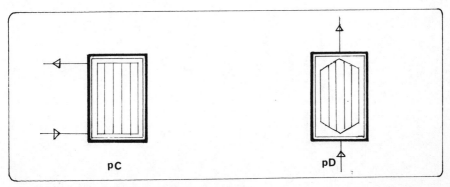

pC            pD

*"pD"* achieves a still more even flow than *"pC"*.

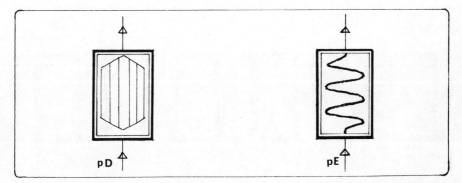

pD            pE

*"pE"*, the serpentine tube has the advantage that the flow velocity will be the same throughout the collector. Velocity will also be higher in *"pE"* and turbulent flow conditions might be achieved thereby increasing increased heat transfer between absorber and water.

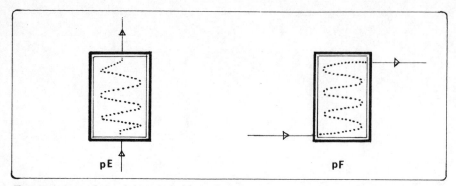

pE

pF

The location of the inlets and the outlets on a serpentine tube absorber make no difference to the performance. When this might be important is if there was not a continuous drop in gradient between the outlet and the inlet. In such a case draining would be difficult.

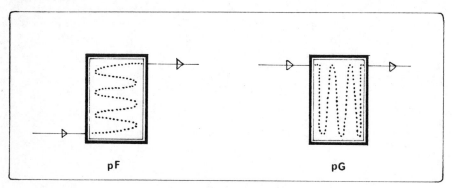

pF

pG

"pF" is best because "pG" will not drain-down easily.

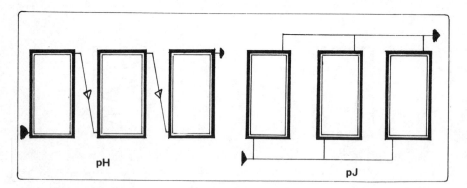

pH

pJ

"pJ" is better than "pH". Being connected 'in parallel' it will operate at lower temperatures and hence at higher efficiencies than "pH" which is in series. If high temperatures are specifically desired however "pH" is the best choice.

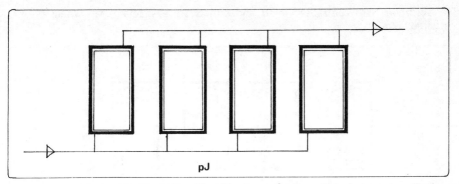

pJ

"*pK*" is better than "*pJ*" which will suffer from the flow disturbances described as the 'MacGregor effect'.

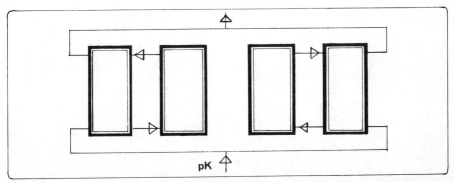

pK

"*pL*" is better than "*pK*" because it has an equally balanced circulation pattern but it has less exposed pipework which means lower heat losses.

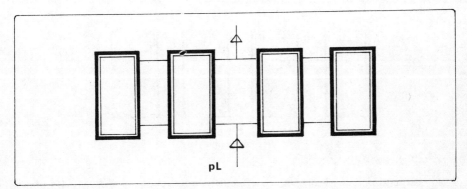

pL

If absorbers with centralised inlets and outlets are available *"pM"* would be the desired layout.

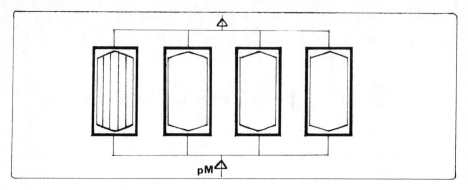

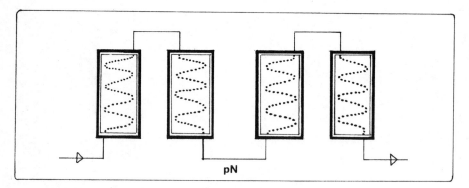

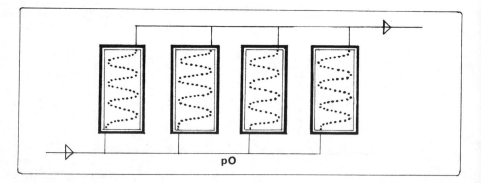

With serpentine type absorbers, the effect of varying pressure differentials is likely to be overshadowed by the frictional resistance between the headers. Thus they can be connected in parallel between headers with outlets at diagonally opposite corners as in *"pO"*. This will be more efficient than *"pN"* because the latter, connected in series, will operate at higher temperatures.

# 11

# Mounting Solar Collectors

## Location

The primary requirement in choosing a location for the collectors is that it should receive as much radiation as possible. They should face no more than 30 degrees away from due south. If you are not sure exactly which direction south is, find out the times of sunrise and sunset, and work out the hour midway between them. This is the solar noon and at that time of day the sun itself will show you due south. Avoid any obstructions, like trees or buildings, which might cast shadows for a large part of the day. Raised areas, like roof tops are usually less overshadowed than walls or gardens.

Having picked out the sunniest areas one should then consider the house's plumbing layout. The longer the connecting pipes between collectors and hot water taps, the more heat will be lost on the way. Ideally collectors, storage tanks and taps will be close together.

Finally, in remote situations the type of circulation desired may be important. If there is no electricity supply to run a pump, the system must be planned to operate with gravity circulation. As explained in chapter 5, this means that the collectors must be mounted below the solar storage tank. Single storey extensions, balconys, walls and ground level locations may be most appropriate in such cases.

## Appearance of Collectors

Solar collectors need not greatly affect the appearance of a building. In some cases they can even enhance it. They have become an eyesore in some instances where frames are built, jutting out from the roof or walls with no relationship to the existing elements of the buildings. Where possible the collectors should be installed so that they line up with any strong lines formed by windows, doors or projections in the building facade.

11.1 Locating collectors to line up with existing elements in the building facade gives a more integrated appearance.

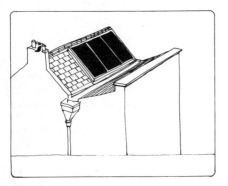

11.2 Collectors mounted on gully roof away from chimneys which might cause overshadowing.

11.3 Extensions form ideal mounts for thermosyphoning collectors. Care is required to avoid overshadowing.

## Mounting on Pitched Roofs

Solar collectors can be inset mounted so as to replace a conventional roof covering or can be surface mounted above the existing slates or tiles.

When the collector is to be an integral part of the roof, the existing tiles or slates must be removed and a supporting framework built above the rafters. (Cutting or notching the rafters themselves is unnecessary and should be avoided as this will weaken the roof structure.) The framework is flashed with zinc or lead at the edges and after the absorber plates are plumbed in, glazing adds the final weatherproofing.

Surface mounting the collectors can be a much quicker operation. In this case metal angles are bolted through the roof and the collectors secured, top and bottom, between them. With this type of mounting, the collectors should have a durable casing such as fibre glass or aluminium. A timber cased collector could be surface mounted but it will be necessary to repaint the casing about every two years in order to maintain the weatherproof finish.

In some situations, where the roof span is short and bounded by parapet walls it is possible to fix a timber frame between the walls and thereby anchor the collectors without actually puncturing the existing roof.

## Safety and Access

The majority of accidents in the building trade occur when people are working with ladders. This does not mean you have to go rushing off to your local builder when it comes to roof work. But it does mean you ought to develop the necessary confidence by careful consideration of how to arrange access to your particular roof.

Extending ladders or scaffolding are the obvious means of access and climbing boards, or roof ladders, which have a curved lip at the top for hooking over the roof ridge, are necessary for actual working on the roof.

*Scaffold*

Scaffolding has the advantage of pro-viding a working platform at roof level. Scaffold towers can be hired from builders and are easily erected as they come in component form and each one simply slots into the next. They usually have wheels on their base and these should be locked before com-mencing work on the tower. The tower should also be tied to the building you are working on.

*Numbers*

Two people is probably the best number for working on the roof. More than that can be distractive and may get in each other's way.

*Ladders*

If you are using ladders then it is best to have two sets, one on each side of the area being worked on. The ladders should be pitched against the building

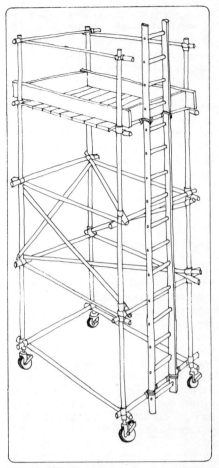

11.4 Scaffolding tower — note ladder tied top and bottom.

so that the base is away from the wall by a quarter of the height it is reaching. The ladder should be secured at the top and the bottom. If there are no existing objects against which it can be secured, screw a ring bolt into the woodwork under the eaves, and drive stakes into the ground to provide anchors for tieing the ladder down. If the ground is con-crete, use sandbags at the ladder base.

Do not lean the ladder on plastic gutter pipes. If the eaves project by a large amount, a stay can be hired which hooks over the top of the ladder and holds it out from the wall.

The most important thing is your mental attitude. Do not be in a hurry. Remember where you are. Concentrate on the job. If you find that impossible, then maybe you should call in your local builder.

### Inset Mounting

Setting the collectors into the roof is probably the most economic method of mounting for anyone building their own system. All exposed parts of the frame work encasing the absorber plates are covered with metal flashing hence the cheapest softwood can be used. Furthermore second hand slates are in demand these days and if removed in good condition, can be sold for as much as 10p. each. A collector of 4sq. metres would replace about £16 worth.

The major disadvantages of inset mounting are the increased time spent working on the roof top and the need for good weather when carrying out the installation.

### Inset Mounting Procedure

The first step is to mark with chalk the intended location of the collectors. General hints about choosing a location are given above. Considering the practical point of view, installation will be easier if they are sited away from the roof edges to avoid disturbing the existing flashing. Access for making pipe connections inside the roof will be facilitated if the collectors are a few feet above the eaves. Remember however that you must have space inside the roof to mount an expansion/feed tank above the top of the collectors.

The lower edge of the collector casing should be positioned about 75 mm (3") above a roof batten i.e. about 150 mm (6") above the lower edge of a row of slates. It will save some slate cutting if it can also be arranged that the upper edge is about 75 mm (3") below the lower edge of a row of slates.

When mounting the collectors in a horizontal bank, the short dimension will be equal to the height of the absorber plate plus 25 mm (1") for tolerance, plus 100 mm (4") for the casing frame members. The length will be the total length of all the absorbers plus an allowance of 60 mm (2½") between them for fittings and at each extremity, plus 100 mm (4") for the framework.

### Removing the Roof Covering

Before beginning to remove the roof covering, it is advisable to have tarpaulins or at least sheets of heavy grade polythene on hand in case of rain. Otherwise you may well find yourself clambering about on a wet slippy roof hurrying to finish the job before the ceiling is soaked through.

Slates and tiles come in many sizes and forms. Slates are always nailed onto battens laid across the rafters, with two rails to each slate. Most tiles have a lip on the upper edge which hooks over the

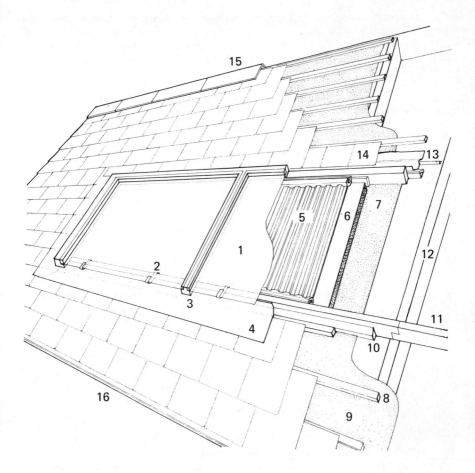

**11.5 Inset Solar Collector**

1) glass
2) lead strap
3) glazing bar
4) flashing
5) absorber (radiator)
6) insulation
7) waterproof membrane
8) roof battens

9) roofing felt
10) fillet
11) casing framework
12) roof rafter
13) kink in flashing
14) half slate
15) Ridge tiles
16) eaves of roof

*11.6 Removing the roof covering from the ridge downwards.*

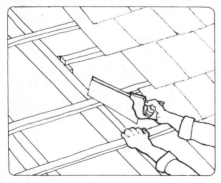

*11.7 Battens are removed from the intended collector area. They are sawn off flush with rafter edge and then supported as shown by a short length of timber nailed to the side of the rafter.*

batten. In some cases they are also nailed every third course. The particular fixing will depend on local weather conditions and the tile design.

If the upper edge of the collectors is close to the ridge of the roof it may be worthwhile removing the ridge tiles by chipping out their mortared seating with a cold chisel. This will expose the nail fixings at the upper edge of the top row of slates which will be half slates. Because slates are laid to overlap the vertical joints of the course below, it is simplest to work diagonally downwards towards the lower corners of the chalked collector, removing the slates course by course. If the collector is more than about three feet from the ridge, this will mean removing a large number of slates and you may prefer to start at the second row wholly above the upper edge of the collector. This is possible with the help of a tool called a ripper — a long handle topped with a flat blade which slips between the courses and hooks round the nails holding down the lower course. The slates which are taken off should be stored, on edge, in a safe place.

When the required area has been cleared, it may be necessary to relocate the collectors slightly by extending its length so that both end frame members can be fixed directly over rafters. The final location being marked, one should clear away the battens which cross the area intended for the collectors. Using a tenon saw, cut them flush with the outside edge of the rafters which will support the frame. A strip of batten should be nailed to the side of the rafters to support the cut ends of the remaining battens.

If the roof has a layer of felt beneath the slates it is advisable to keep this intact. It will serve as a failsafe if leaks

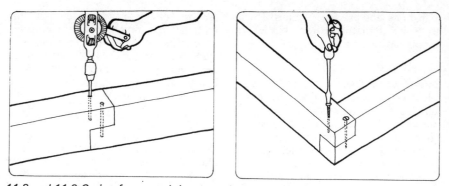

*11.8 and 11.9 Casing framework is cut to size using lap joints.*

ever developed. During construction however it can be very helpful to have easy access to the attic. A small window can be cut within the area to be occupied by the absorbers, and this can later be patched with a larger sheet of felt pasted on with mastic on the overlaps.

### Casing

The casing consists of 50 x 100 mm (2" x 4") timber. It should be cut to size with lap joints at ground level and assembled on the roof. It can be fixed with 100 mm (4") nails driven in diagonally from both sides. The frame should be surrounded by a triangular section fillet cut from a 75 x 50 mm (3" x 2").

*11.10 Framework is nailed to rafters.*

Inside the frame, small blocks should be screwed 25 mm (1") below the upper surface. These will support the absorber plate.

### Installing Absorbers

The absorber plates should now be laid upon the blocks in the casing. When working on a scaffold tower, they can be hauled up with ropes. Alternatively two people should take an end each and lift them between the two ladders. Once installed they should be connected in the most appropriate pattern shown in chapter 10.

If the absorber inlets and outlets are copper pipes they can best be joined

*11.11 Absorber support blocks fixed.*

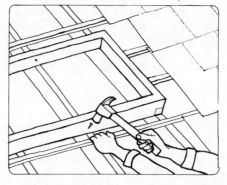

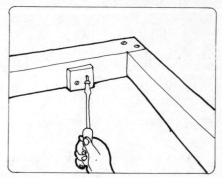

with brass compression fittings. If you are using radiator panels, black iron unions can be used.

Care must be taken with the hot water outlet where it goes through the roof to join the flow pipe leading to the solar storage tank. This outlet pipe should rise until it meets the vent pipe which discharges over the expansion tank. If you simply took an elbow out of the absorber and led the pipe downwards into the attic, there would be a danger of an airlock forming at this point and reducing or blocking the flow rate.

It will be necessary therefore, except on steeply pitched roofs, to notch the underside of the casing framework in order to allow this slight upward gradient in the outlet pipe.

On shallow pitched roofs this may be impossible and space inside the casing should be allowed for an automatic air vent to be fitted on the outlet at its highest point.

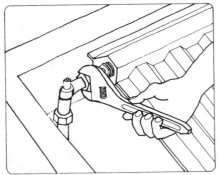

*11.12 Collector outlet (flow pipe) must be vented at its highest point.*

### Flashing

A roll of flashing material at least 250 mm (10″) wide is now required to weather-seal the casing. Lead and zinc are the materials in common use for this purpose. Zinc is about 20 per cent cheaper than lead but not so easily formed round bends and requires soldering at corners. Unless you are experienced with zinc then, lead is the best choice. It comes in several thicknesses and for this application 1.8 or 2.24 mm (4 or 5 lbs per sq. ft) is recommended. A much cheaper, but also shorter lived material would be aluminium faced bitumin roll. This is easily moulded round corners and is self-adhesive.

Before fixing the flashing the slates should be replaced around the lower edge and sides of the collector. The flashing should then be nailed with galvanised clout nails to the upper surface along the lower edge of the frame. This sheet should project at the edges where it will be overlapped by the side flashing. Flashing on the upper edge is fixed last. This forms a kind of gutter for the roof above. If it is a steep roof the gutter must be deeper. In such cases a wider strip must be used and this should extend over the glazing bar. A triangular fillet the same depth as the slate battens should be nailed to the rafters some 75 mm (3″) above the casing's upper edge. This forms a barrier to prevent rain washing back and over the flashing, which should be moulded over it.

The flashing should not be fixed in contact with, or in a position where it would receive the run-off water from any copper piping or sheet. Otherwise the flashing would corrode.

### Replacing the Roof Covering

The slates above the collector can now be replaced. A row of half slates should

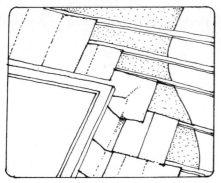

be placed to cover the flashing along the top of the collector. These can be cut by scoring the slates then chopping along the edge of the mark with a flat bladed trowel.

Nail holes can be made along their upper edge by laying the slate on a flat even surface and hammering a blunt nail through, or by using a masonry bit in a slow speed drill.

*11.13 Upper strip of flashing laid to overlap side strips. Slates above collector can then be replaced.*

The row of half slates must be covered by a row of whole slates, nailed to the same batten. Remember that the vertical joints must be offset, so that the

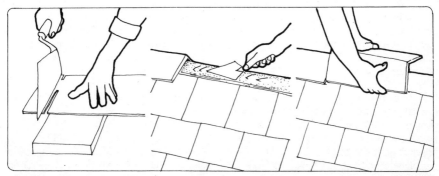

*11.14 Cutting slates and replacing ridge tiles bedded in mortar.*

upper slates cover the joints of the lower row.

*11.15 Fixing the final slates in places with lead straps nailed to the battens.*

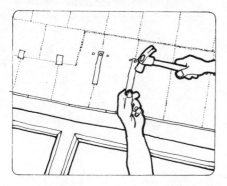

If you removed only two rows above the collector by using the ripper, the final row will have to be secured by lead or zinc straps as the batten into which the slates were formerly nailed is now covered by the surrounding slates. These straps, 9" x 1" (225 mm. x 25 mm.) should be nailed through the vertical joints between the slates you have just fixed and into the same batten. Carefully slide the remaining slates into position above the straps then fold the straps over their lower edge to prevent them slipping out.

## Glazing

The glazing system described here makes use of timber glazing bars which would require access to a bench circular saw for its construction. Alternative glazing details are described in ch. 12 which can be made up using common hand tools.

Glazing bars are fixed above the top frame member and both side members. In addition, if the collector is larger than about 1500 mm x 600 mm (5ft. x 2ft.) you will need more than one sheet of glass and extra vertical supports (mullions) will be required. These mullions will cast a small shadow on the absorber below and should if possible be positioned above joints between absorber panels. Along the lower edge the glass projects over the frame to carry rainwater clear of the casing. It is held in place by lead straps, spaced at 600 mm (2ft.) centres. The straps must be nailed to the inside of the lower frame member before the glass is in position. Also required is a strip of 9 mm (3/8") wood and mastic to fill the gap between the glass and the lower frame member.

The glazing bars are cut from a 63 mm x 36 mm (2½" x 1¼") section of hardwood, preferably teak for durability. A rebate 20 mm x 20 mm (¾" x ¾") is cut to receive the glass, and a small groove is gouged along the complete length to form a drip where the glazing bar project over the frame members.

The glazing bars should now be screwed down on the frame, with screw heads being countersunk and covered with

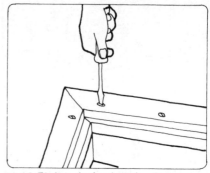

*11.16 Fixing glazing bars in place.*

mastic. If you are using Dutch lights you will have spaced the bars to suit the standard dimensions (see ch. 12). Otherwise you should now measure up the glazing bars to determine the size of glass panes required. Allow 3 mm (1/8") for tolerance and thermal expansion on each side. Allow also for the glass to project 37 mm (1½") over the lower edge of the frame to carry rain water clear of the casing.

Great care is required when handling glass on the roof both to avoid personal injury and damage to the material. The glass should be bedded in a silicon sealing mastic. The seal should then be covered by a 12 mm x 12 mm (½" x

*11.17 Sealing mastic for bedding glass.*

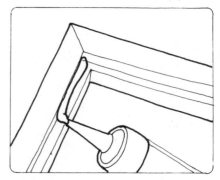

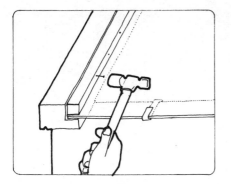

The mounting is now complete and you can get your feet back onto solid ground and admire your new sun-catcher roof.

**Surface Mounting**

½") beading which is also bedded in mastic and nailed into the glazing bar.

Manufacturers who produce solar collectors with durable casings often recommend a much quicker form of surface mounting. Many kits are fixed by bolting to a metal angle which is clamped above the slates or tiles by J-

**11.19 Surface Mounted Collectors**
*N.B. The location of collector inlet and outlet shown here is that most commonly adopted. As explained in ch.10 however, alternative connection layouts might improve the pattern of water flow through the panels.*

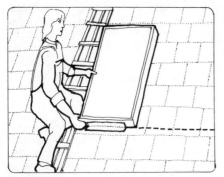

11.20 Mark out intended location of solar collectors.

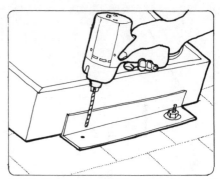

11.21 Holes are drilled in the roof covering and packed with silicone sealant. Corresponding holes are drilled in the collector mounting angle.

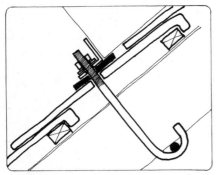

11.22 Cross-section showing J-bolt hooked round steel rod. Note felt packing between collector angle and tile. Don't forget to seal the hole.

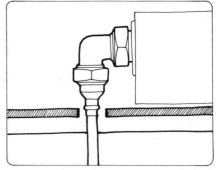

11.23 Collector inlet pipe passes through hole drilled in tile.

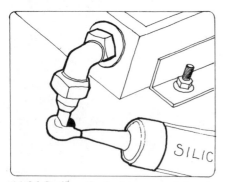

11.24 Sealing the hole prepared for the inlet pipe.

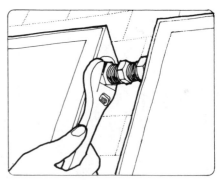

11.25 Connecting the collector panels with straight compression couples.

100

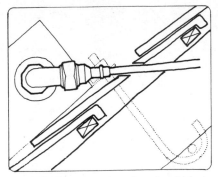

*11.26 Collector outlet pipe should have a rising gradient until it reaches junction with its vent pipe.*

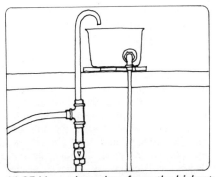

*11.27 Vent pipe taken from the highest point on the collector outlet, discharges over expansion tank. Note one-way valve on collector outlet pipe and solar primary feed pipe coming from tank.*

*11.28 Flat roof collector mounting.*

bolts which pass through them and hook round a 12 mm (½") diameter metal rod on the underside of the roof rafters. Holes for the J-bolts are drilled with a 7 mm (5/16") masonry bit and a weatherseal is achieved by packing the hole with silicone sealant.

When surface mounting collectors, it is important to realise that you are not concerned simply with preventing the collector slipping off the roof. The main threat is the wind, not gravity. In the most exposed conditions, where the collectors are mounted on the lee side of a double pitched roof, the upward acting suction force might be as high as 20 lbs/sq. ft. (900 Newtons/sq.ft.).

The collectors must therefore be fixed both top and bottom with at least four bolts or screws and these should be ¼" (6 mm.) diameter. They should be fixed near each corner in order to prevent any rocking movement during windy weather.

**Flat Roofs**

On a flat roof a supporting framework is required in order to tilt the collectors towards south. Several companies sell purpose made metal frames to support their solar panels. A simple triangulated frame can be constructed from timber. The wood should be treated with preservative and should not be laid direct on to the roof surface in such a way that it might be unventilated. Thus the base plates of the structure should be fixed on spacer pads. These pads can be formed from roofing felt, nylon, lead or any material which will not corrode in wet conditions.

101

Where the roof is bitumin or felt laid on boarding, the fixing can take the form of No. 12 screws with a mastic bedding. The mastic chosen must be compatible with the roofing material. Where the roof is constructed with concrete, something like Rawlbolt fixing will be required.

In choosing the collector location, factors such as overshadowing are of primary importance. But it would be wise where possible to avoid those areas of roof where puddles form after rain.

**Vertical Mounting on Walls**

*11.30 Wall mounted collectors fixed on a simple triangulated frame.*

*11.29 Wall mounted collectors integrated with the building facade.*

In some situations a wall may be the only available location for solar collectors. This will undoubtedly have a greater effect on the appearance of the house. This need not be a detrimental effect. Indeed, they can add both to the appearance and the function of the building. Mounted over an entrance they can form a sheltering roof. Over a window they can form an awning which would provide shading in summer.

In these applications the collectors would be mounted at a tilt. Collectors have sometimes been mounted vertically. Such steep angles of tilt can perform well on sunny days in the middle of winter. Taken over the whole year however, you would need almost twice the number of vertical collectors as 45 degree collectors in order to yield similar quantities of energy. Projecting the foot of the collector away from the wall by even a small amount can cause a dramatic improvement · in performance.

One company uses 1/8" (3.2 mm.) aluminium angle to wall mount their collectors. Another uses a triangular steel web to fix to a central pivot point on their collectors thereby allowing the angle of tilt to be altered.

# 12

# DIY Collector Plans

The simplest solar collector for home construction is a timber cased radiator panel of the type used in central heating systems. Attention should be given in the first instance to the selection of a suitable radiator.

They can be bought from a plumbers merchant for about £19/m². Second hand radiators form what must be by far the cheapest solar absorber. In junk yards, or discarded in neighbours gardens I have found lots of second hand radiators and bought them for about 50 pence each.

The best radiators for solar collectors are those which hold little water, the modern slim profile models. These will have the minimum thermal mass and will therefore react more quickly to short intervals of sunshine. Double panel radiators and the old fashioned segmented radiators are of no use.

The second feature to examine is the connections. The most adaptable radiators are those with a 1" or ¾" threaded hole in each corner. Some radiators have in one corner, only a small hole. This is

for an air vent. It is of no use for circulation connections and should be blocked with a threaded iron plug. The most difficult radiators to use are those with only two holes — both on the same header. These should be used, if at all, only on pumped systems with high flow velocities. Refer to chapter 10 before deciding upon the positions of your inlet and outlet.

If you have access to brazing equipment, the position of the existing connection holes is not so critical as a hole can be easily drilled in the ends of most radiators and a 150 mm (6 inch) length of copper pipe brazed into it.

When choosing the dimensions of the radiators one must take account of the dimensions of the space in which they are to be mounted and the ease of handling which is desired if they are to be roof mounted. Once encased and glazed, any radiator larger than 600 x 600 (2 ft x 2 ft) will require more than one person to carry it. Large radiators have the advantage of economy. Fewer plumbing connectors are required and one large case requires less wood than several smaller ones. From my own

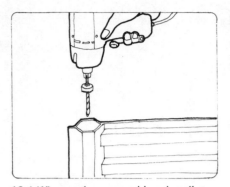

12.1 When using second-hand radiators which lack connecting holes in the required location, a new hole can be drilled and a copper pipe brazed in place.

12.2 Brazing a copper pipe to a hole drilled in a radiator. N.B. Most radiators have conveniently placed threaded holes and when buying new panels you can specify your requirements. Brazing equipment therefore is NOT a necessary requirement for making solar collectors.

experience of working on rooftops though, I would not choose to handle anything larger than a 1500 x 600 mm (5 ft x 2 ft) panel, except on a sheltered flat roof.

Another point to be borne in mind when choosing a radiator is the dimensions of convenient casing materials, particularly glazing. Horticultural glass for example is produced in standard sizes and if you have a supplier nearby, usually nurserymen, it is worth trying to tailor your collector design to suit these dimensions. This will cut the costs as horticultural glass is about half the price of window glass and usually has identical properties of energy transmission. The standard sizes are shown in the table below. The Dutch Light categories are the most suitable for solar collectors.

| 3mm thick | Dutch Lights 3mm or 4mm thick | Dutch Lights 4mm thick |
|---|---|---|
| 457x305 | 1410x730 | 900x800 |
| 457x406 | 1422x730 | 1000x800 |
| 508x457 | 1613x730 | |
| 610x305 | 1651x730 | |
| 610x356 | 1676x730 | |
| 610x406 | 1778x730 | |
| 610x457 | | |
| 610x610 | | |

Table 12.1 Horticultural Glass — Standard Dimensions (millimetres)

## CASING

Aluminium and fibre-glass are the most common materials used for encasing commercial collectors due to their light weight and durability. If you have had experience in working with these materials, by all means use them. Most people however will find timber much more easily available and easier to work with. And, if it is treated with preservative at the outset, and regularly re-coated or painted, it will endure for many years.

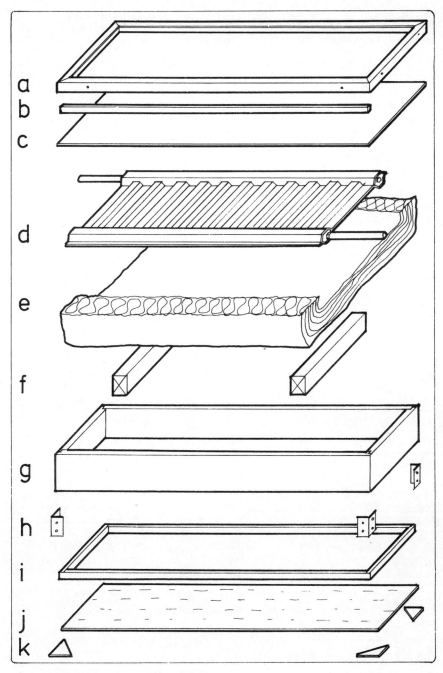

a
b
c
d
e
f
g
h
i
j
k

*12.3 Exploded view of a solar collector.*

| No. in Illus- tration | Component | No. Reqd. | Material | Dimensions Millimetres (inches) |
|---|---|---|---|---|
| a | Glazing Angle | 2 | Aluminium Angle | 1434 x 25 x 25 x 2 (56.3/8 x 1 x 1 x 1/16) |
| | | 2 | | 758 x 25 x 25 x 2 (29¾ x 1 x 1 x 1/16) |
| b | Glazing Angle Spacer | 1 | Softwood Strip | 1430 x 20 x 4 (56¼ x ¾ x 1/8) |
| c | Glazing | 1 | Dutch Light | 1410 x 730 x 4 (55½ x 28¾ x 5/32) |
| d | Absorber Plate | 1 | Central Heating Radiator Panel or Copper Tube-and- Strip Absorber | 1330 x 700 (max.) (52½ x 27½ (max.)) 1380 x 700 (max.) (54¼ x 27½ (max.)) |
| e | Insulation | | Mineral wool or glass fibre quilt | 75mm (3") |
| f | Absorber Support | 2 | Softwood Battens | 700 x 50 x 50 (27½ x 2 x 2) |
| g | Side Panels | 2 | Preferably Hardwood | 1430 x 125 x 25 (56¼ x 5 x 1) |
| | | 2 | | 725 x 125 x 25 (28½ x 5 x 1) |
| h | Corner Angle | 4 | Aluminium Angle | 109 x 50 x 50 x 2 (4.3/8 x 2 x 2 x 1/16) |
| i | Base Seating | 2 | Softwood Strips | 1354 x 25 x 13 (53¼ x 1 x ½) |
| | | 2 | | 700 x 25 x 13 (27½ x 1 x ½) |
| j | Base Board | 1 | Exterior quality Ply or Oil Temp'd Hardboard | 1380 x 700 x 6 (54¼ x 27½ x ¼) |
| k | Corner Bracing | 4 | Exterior quality Plywood (triangles) | 125 x 125 x 6 (5 x 5 x ¼) |

*Component List for DIY Solar Collector.*

## FIXINGS

| Location | No. Reqd. | Item | Size mm | inches |
|---|---|---|---|---|
| Glazing and corner Angles | 26 | Screws — round headed | 25 | 1 |
| Base and corner braces | 26 | Screws | 25 | 1 |
| Base seating, corner joints, glazing angle spacer and absorber locators | 28 | Nails — oval | 30 | 1¼ |
| Absorber supports | 8 | Screws | 50 | 2 |
| Corner joints | — | Wood Glue | | |
| Glass/Aluminium joint | 1 tube | Silicon Sealant (plus primer when recommended by manufacturer) | | |
| All timber | | Timber Preservative | | |

*Table 12.3 Fixings List for DIY Solar Collector.*

N.B. The screws will be exposed to weathering and therefore plated or stainless steel ones should be used.

The lists in tables 12.2 and 12.3 above shows the materials you require to construct a timber casing which would accept a 1410 x 730 mm (55½" x 28¾") Dutch Light and 600 mm (24") high radiator panel up to 1330 mm (52½") in length.

*12.4 Collector base board — component no. "j".*

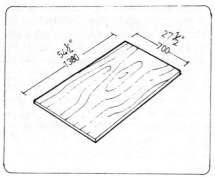

Although a hardwood such as teak is recommended for the side panels, cheaper softwoods and even recycled floorboards have often been used. Remember however that if you do not use boards of the same thickness as those indicated in the table, this will have an affect on the overall dimensions. Always work to the casing's internal dimensions which are critical if the glass is to fit properly. For this reason it is advisable to start by cutting out the base board and then build the box around it.

Mark out the base board's outline care-

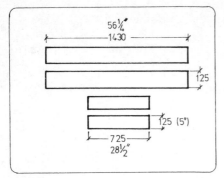

12.5 Side panels — component no. "g".

fully using a straight-edge and a right-angle. Check that it is a true rectangle by ensuring that the diagonals are equal before sawing it out.

The side panels should be formed into a rectangle around the base with lap joints at the corners. These are easily formed by sawing out a 12.5 x 12.5 (½" x ½") square section housing from each end of all four panels. Mark out the waste section clearly using square edge, then cut with a tenon saw, gripping the panel in a vice.

12.6 Marking and sawing out lap joint housings in side panels.

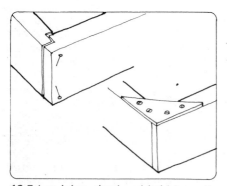

12.7 Lap joint, glued and held by nails, then reinforced with plywood corner bracing (component "k").

Check that the joints fit tightly into each other and that the base fits tightly inside. Now coat the contact surfaces of the joint with glue and follow the manufacturer's instructions regarding drying and setting times. Hold the panels together during the setting period by driving a couple of nails into each joint. These should be dovetailed, i.e. hammered in at opposing angles. You should also fix the corner bracing at this time. This consists of four triangles of plywood screwed into the underside of the side panels where they meet.

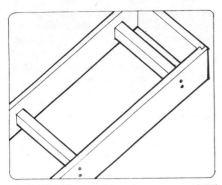

*12.8 Absorber supports (component "f") fixed inside casing.*

The upper side of the frame will support the glass and must therefore be even. Check this using a spirit level and plane down any irregularities.

The absorber supports should now be fitted. They span between the two long side panels and are fixed by screwing through the panels from the outside. They should be situated so as to leave a gap of at least 75 mm (3") in the casing corners to allow for manipulation of the absorber connections. Also allow for the thickness of the absorber, plus an air gap of at least 13 mm (½") between the top of the support and the top of the casing.

Now nail on the base board seating strips so that their lower edge is 75 mm (3") below the top of the absorber supports. Again, the nails should be dovetailed.

The base board can now be screwed on if the assembly is turned upside-down. If the base is going to be exposed to driving rain, it is important to bed the base in a sealing mastic in order to keep the insulation dry.

*12.9 Nailing base board seating strips (component "i") to side panels.*

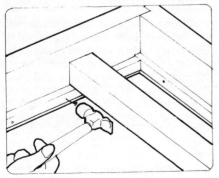

A 75 mm (3") deep layer of mineral wool or glass fibre quilt can now be laid inside the box, turning it up at the edges to reduce heat losses through the sides. These are the insulants usually used for roof insulation and if you have ever used them you will remember that they can cause temporary irritation to the skin. Gloves should therefore be worn and a face mask should also be used to prevent the possible health hazards associated with inhaling the fine dust particles which the insulants give off when handled.

The absorber itself can now be laid inside the casing. Centre it upon its

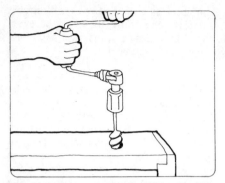

12.10 After carefully marking absorber inlet and outlet positions, drill holes to accommodate connecting pipes.

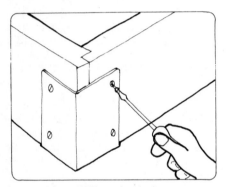

12.11 After drilling pipe holes, screw on corner angles (component "h").

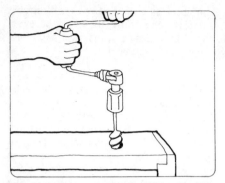

12.2 A ¾" male iron/22 mm compression fitting for connecting copper pipe to radiator.

supports either by notching them, or by driving nails into the supports above and below the absorber. Mark the inlet and outlet positions accurately on the side panels. The holes to accommodate these pipes must then be drilled.

With the inlet and outlet holes prepared, the aluminium corner angles can be screwed in place. These will protect the exposed end grain of the timber and reinforce the framework. For added protection against the weather, they should be bedded in mastic. They should be located flush with the underside of the side panels, but with an allowance at the top for the glazing angle — the angle's internal width minus the glass thickness. The corner angles may need to be sawn or drilled to accommodate the inlet and outlet pipes.

The glazing will be added after the absorber panel is prepared and finally installed.

**Radiator Connections**

Assuming that your radiator has four ¾" threaded holes, you will need two ¾" iron plugs and two straight couples consisting of a ¾" iron at one end, and a 22 mm compression fitting at the other. Screw these into the radiator using PTFE tape or boss-white to seal the joint, and tighten with a spanner.

You should now check for water tightness by closing one compression fitting

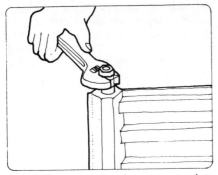

*12.13 Tightening the pipe connector into the radiator.*

with a blanking disc and filling the radiator through the remaining opening. Leave the radiator standing like this for an hour in a dry place above a strip of dry paper. The position of the blanking disc should then be altered and the experiment repeated with the radiator upside-down.

This may seem very time consuming but the aggravation of finding a leak after the radiator has been sealed in its case is worth guarding against, particularly if you are using second-hand radiator panels.

## Painting Radiators

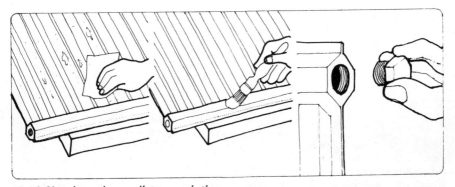

*12.14 Cleaning the radiator; painting the upper surface; closing unwanted holes with a threaded iron plug-wound with PTFE sealing tape.*

Once convinced that there are no leaks, the panels can be painted. Clean any grease from their surface and dry them down before starting. Any rust which has formed on the radiators should be removed with a wire brush.

Radiator panels are usually coated with a thick layer of light coloured paint. This will resist to some extent the flow of heat from the surface into the interior. Because of this it is sometimes recommended that the upper surface be paint stripped with a blow lamp. This is a laborious process and I have never

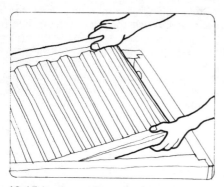

12.15 Laying radiator in the casing.

followed that advice. The existing coating, if in good condition, forms a good barrier to surface corrosion, and forms a non-creep base for the coat of black paint which will be applied on the upper side of the panel.

If the panels are stripped down, a self-etching rust inhibiting primer should be applied before the matt black paint. Wash type primers are best as they form only a thin layer.

I have always used ordinary blackboard paint for absorbing black surface and as yet I have not had any difficulties with blistering or flaking. This might well become a problem if the panels are left in the sun, under glass without water flowing through them. Special paints are made for application on hot metal surfaces such as old-fashioned cast-iron stoves and motor-bike exhaust pipes. These are of course more expensive than ordinary paints. Anyone looking for a really deluxe coating might like to try Nextel, the coating produced by 3M Co. The absorber, by the way, only needs to be black on the upper surface.

12.16 Short lengths of copper pipe should be inserted through the casing. Connecting pipework can later be soldered, compressed or clipped to the projecting ends.

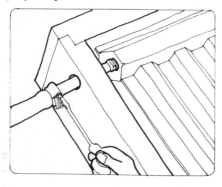

**Fixing Inlets and Outlets**

Once painted the radiator can be laid in the casing. Short lengths of 22 mm copper tube should now be inserted through the holes drilled in the casing. They should be engaged in the compression fittings and the joint made fast by tightening the fitting as described in chapter 8. The copper pipes should project at least 50 mm (2") outside the casing, and longer if you intend to use capilliary copper fittings.

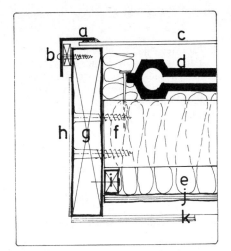

*12.17 Cross-sectional view of solar collector.*
a) glazing angle
b) glazing angle spacer strip (on bottom side only).
c) glass
d) absorber plate
e) insulation
f) absorber support
g) side panel of casing
h) corner angle
i) base seating
j) base board
k) plywood corner bracing

With the absorber fitted inside the casing, the glazing can now be added. The cheapest method is traditional putty glazing. Under conditions of high temperature and direct sun however, this will need frequent replacement. A silicone sealant is more expensive but much longer lasting. If the sealant is also covered by the metal angles which retain the glass, its life may be further prolonged.

Begin by nailing the glazing angle spacer strip along the upper external edge of the bottom side panel. It should protrude slightly above the upper edge of the panel side, no more than 2 mm (1/8''), to act as a stop should the glass ever slide downwards. The spacer strip's main function is to ensure that the aluminium glazing angle along the lower side projects out from the timber casing so as to shed the rain water which runs across the collector.

The glass can now be laid in place, equally lapping the case at each side. Check that the casing is providing a level support all round the glass otherwise it may be cracked when the glazing angles are screwed down.

A thin line of silicone sealant should be extruded all round the perimeter of the upper surface of the glass where it will be overlapped by the glazing angle. Care should be taken to prevent the silicone running over the side of the glass and coming into contact with the glass/timber joint. This would cause the glass to adhere to the timber and it would be difficult to remove the glass should this ever be necessary in the future.

The right-angle section aluminium which forms the glazing angle should be cut to form a chamfered joint at the corners.

113

To do this accurately you need a 45 degree angle. Use this to scribe a guide line on what will be the horizontal surface of the glazing angle. The angle should be cut with a small hacksaw whilst held firmly by clamps or in a vice. The vertical side of the angle can be chamfered with a file.

Screw holes should be drilled in the vertical side of the angle at the middle and near each end. Make sure you do not drill through the horizontal surfaces, the ends of which you have just sawn off at 45 degrees. It is sometimes advised to use a few drops of paraffin or turpentine as a lubricant when drilling through aluminium.

Before laying the aluminium angle in place, it may be necessary to paint their inner surfaces with a special primer to ensure adhesion with the silicon sealant. Follow the instructions of the sealant manufacturers in this matter.

With the glazing angles in place, have someone press them down, lightly but firmly, whilst you drill into the casing through the screw holes prepared in the angle. Finally, fix with round-headed screws, and the solar collector is complete.

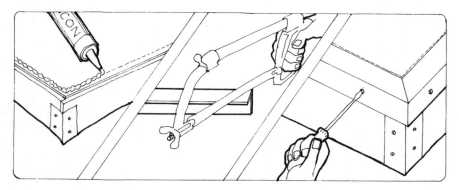

*12.18 Extruding a thin line of silicone sealant along the perimeter of the glass; cutting a chamfer in the glazing angle; screwing the angle to the side panels.*

## De-Scaling Old Radiators

If you have used a second hand radiator to make a solar collector, it may be worthwhile giving it a chemical bath to clean out the layers of rust and other deposits which might have formed inside the panel. These layers might be like the scale which forms inside kettles and saucepans in hard water regions. They have the effect of resisting the flow of heat through the radiator, from the absorbent black surface to the circulating water inside.

Builders' merchants sell proprietary preparations for carrying out descaling and their instructions should be followed carefully. If these are unobtainable, descaling can be carried out with a simple solution of dilute acid. A mixture of 10% hydrochloric acid will do the job. I have heard that even vinegar can be used. Weak acids will of course require considerable time to loosen deposits so the radiator should be filled and left for about 12 hours. It should then be washed out with water and then refilled with water. It should be kept filled now until its installation as acid cleaned steel surface will rapidly rust if air is allowed to enter by draining the panel.

Remember of course that caution is important when handling acid. Wear thick old clothes, avoid splashing, immediately wash with clean water if any concenttraded acid does come in contact with your skin and take particular care to avoid contact with your eyes.

## DIY Copper Absorbers

A copper absorber can be made which will last longer and perform more efficiently than a radiator. Its material cost will be about the same as that of a new radiator panel but it will obviously require more work in its construction.

As described in chapter 4, there are two types of copper tube-and-sheet absorbers: grids of parallel pipes between two headers and serpentine or zig-zag pipe configuration. If the system is to operate without a pump, the grid type is best. The serpentine form however is much cheaper and quicker to fabricate as it has far fewer joints. The number of tee fittings required per unit of absorber area in the grid type can be minimised by dimensioning it so that distance between the two headers is as long as possible.

## Serpentine Tube

In forming the serpentine, or zig-zag absorber tube, one of the greatest difficulties is bending the tube in such a way that we are left with a level piece of pipework which can be easily bonded to the flat copper sheet which absorbs the radiation which falls between the coils of tube. The more bends we make, the more difficult it is to keep it level.

The design outlined here is for an absorber which will fit the casing already described and is based upon two 3 m (9'11") lengths of 15 mm copper pipe. This is the standard length in which copper tube is stocked so there will be no unnecessary wastage or extra charges for cutting tube to size. Furthermore, each tube is given only one bend then the two are joined together with two capilliary elbow fittings. Using this method it is much easier for the beginner

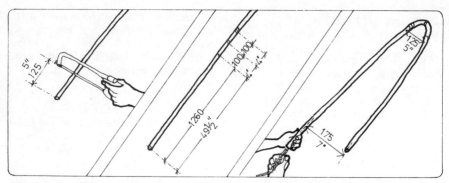

*12.19 Cut a 125 mm length from the end of the pipe; move bending zone; removing bending spring after forming U-bend.*

to achieve a level layer of zig-zag pipe.

Begin by cutting off a 125 mm (5″) length from the end of one pipe. Keep the short piece in a safe place as we will be using it later.

Now measure and mark three points on the pipe, at 1260 mm, 1360 mm and at 1460 mm from the end, 4′ 1½″, 4′ 5½″ and 4′ 9½″). These will be the start, middle and finishing points of the 180 degree, U-bend you must make in the pipe. The bending is done with a plumbers' spring and can be done manually or with a machine and discussed in chapter 8.

The bend should actually be slightly less than 180 degrees so that the arms of the U-bend diverge slightly rather than being parallel. Aim for a spacing of 125 mm (5″) between the centres of the tube at the points marked for the start and finishing of the bend, and a spacing of 175 mm (7″) at the ends of the tube. This is done so as to avoid having rows of parallel, horizontal tubes which would be difficult to drain once the collector is mounted in a fixed position.

You will notice that one of the arms is somewhat longer than the other. This is intentional. The longer arm will protrude from the casing to serve as the inlet or outlet.

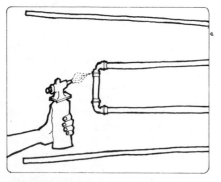

*12.20 Soldering together two bent lengths of copper pipe using capilliary elbows.*

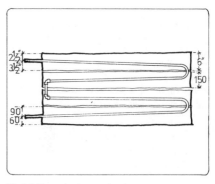

*12.21 Serpentine tube located over four lengths of copper strip.*

The second length of pipe should be bent in the same way as the first and then both lengths should be laid side-by-side on a level surface with the shortest arms closest to each other. They can now be joined using the 125 mm (5") length, which was cut off in the beginning, together with two capilliary elbow fittings. These should be bonded with a blowtorch as described in chapter 8.

With the waterways prepared, we can bring our attention to the copper backing sheet. It is worth noting first of all that we are using copper because it is the easiest material with which we can make a good thermal bond with the copper tube. This bond is a critical factor affecting the efficiency of tube-and-sheet absorber plates. We make it as good as possible by folding the sheet tightly round the tube and running solder between the two.

Copper sheet is most commonly available in rolls 150 mm (6") or 300 mm (12") wide. Narrow strips like this are quite convenient for solar absorber construction. When working with wider sheets it is difficult to form tight-fitting folds around all the tubes. Unless you have made a large press which forms the whole sheet in one go, there is a creep effect which tends to slacken the folds. You will perhaps notice also that the 150 mm (6") width matches our average tube spacing.

There are three types of copper sheet: soft, half-hard and hard. "Hard" should not be used as it is extremely difficult to fold it around the tubes. "Half-hard" and "soft" are much easier to work with but "soft" is very soft and therefore for the sake of a neat undented appearance, most manufacturers use "half-hard".

For economy and ease of folding, choose a thin sheet; 0.25 mm 10/1000" or 33 SWG) thickness is ideal.

You will need 5.4 metres (17' 10") of the 150 mm wide (6") copper strip. This must be cut into four equal lengths of 1350 mm (53½"). Lay these strips across the zig-zag tube as shown in fig. 12.21. A semi-circular groove can be formed in the strip by forcing it around the tube, using a wooden block as a former.

To do this, select a block about 100 x 100 x 75 mm (4" x 4" x 3"). Draw a line around the block about 25 mm (1")

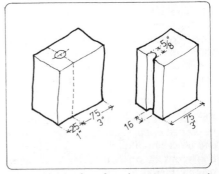

12.22 Block for forming copper strip round 15 mm tube.

above its base. Then drill a 16 mm (5/8") diameter hole right through the block, centred on this line. If you now saw off the lower 25 mm (1") using the line you have drawn as a guide, you should be left with a 100 x 75 x 75 mm (4"·x 3" x 3") block with a semicircular groove along its base, which can wrap the copper sheet tightly around 15 mm tubes.

If you prepare three sets of wooden blocks, and have access to three large clamps, the easiest way to form the copper strip is to clamp the tube to the strip at each end between the blocks. The third set of blocks should then be clamped first in the centre and then in progressive steps in each direction until a tightly fitting groove is formed along the complete length of the strip.

To complete the tube-and-strip bond, they must be soldered. Clean down the contacting surfaces with wire wool, smear with flux, heat with a blowtorch and add solder in a manner similar to that described in chapter 8 for soldering end-feed fittings. The soldering can be made easier by using solder paint. This is a thick fluid which is painted on both contacting surfaces after cleaning. It combines flux and solder and after the surfaces are brought together they simply require heating with a blowtorch to complete their bond. Additional solder should be added where there are gaps between the contacting surfaces.

For the best results, the soldering should be carried out in short stages with clamps and blocks holding the tube and sheet in tight contact on both sides of the area being soldered.

The assembly should then be cleaned and painted with a wash-type self-etching primer before being coated with a matt black absorbing finish.

**Parallel Grid Absorbers**

If you are intending to build a thermo-syphoning solar system, it would be better to make the absorber from a grid of parallel tubes between two header pipes.

Cut four 1270 mm (50") lengths of 15 mm. tube and four equally long pieces of 150 mm. (6") wide copper strip. Cut a notch in the centre of each end of the strip to accommodate the tee fitting into which the tube end will be placed. Now form the copper strip round the tube and solder them together as described in the section on serpentine tube absorber construction.

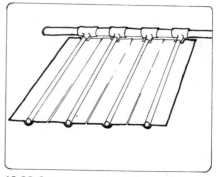

12.23 Cut-away view of parallel grid absorber.

You will need eight 28 x 28 x 15 mm reducing tees plus four short lengths, and six still shorter lengths, of 28 mm pipe to form the header pipes which connect the tube-and-strips at each end. After cleaning and fluxing, assemble the whole grid on a flat surface with the tees projecting over the edge so as to be accessible to the blowtorch flame. Solder the whole header at one time. The tees may slip out of alignment if assembled individually. It is advisable to lay a wet rag across the upper edge of the tube and strip joints to prevent the solder in them melting when heating the tee fittings.

118

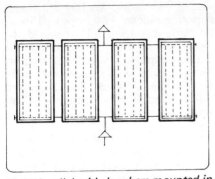

## MOUNTING DIY COPPER ABSORBERS

The grid absorbers should be mounted with their 15 mm tubes running from top to bottom to facilitate thermosyphoning circulation through them. The serpentine absorbers, on the other hand, should be set with their tubes running from side to side, i.e., near horizontal, so that it is possible to drain the collectors without having water trapped in the zig-zag coils.

*12.24 Parallel grid absorbers mounted in parallel with the tubes running vertically to facilitate gravity circulation and to discourage the formation of air blockages.*

For detailed notes on the effects of different connection patterns, see chapter 10.

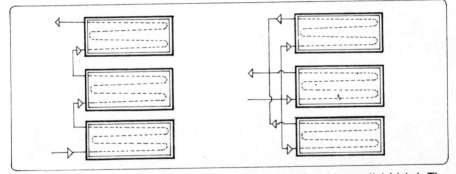

*12.25 Serpentine tube absorbers mounted in series (left) and in parallel (right). The tubes run near horizontal enabling the collectors to be drained down.*

## HOW MANY COLLECTORS?

If you have followed the instructions given in this chapter, the solar collector you have made will have an absorber plate area of just over 0.8 sq. metres (8½ sq. ft.) If you are going to use it in a domestic solar water heating system, you will need more than one. The number required depends upon the amount of hot water you use.

The average household consumes 160 litres/day (35 UK galls/day) that represents 50-60 litres per person (11-13 UK galls/person). Average figures however can often be misleading as the amounts used by different individuals varies tremendously. To play safe, engineers often design hot water systems to supply a generous 100 litres of hot water per day for each person. (20 gallons/person/day) with the recommended minimum being 70 litres/person/day (15½ UK galls/person/day).

Once you have decided what your hot water requirements are likely to be, there is a simple ratio to help you calculate the amount of solar absorber area that you need: allow 1 sq.m. of absorber for each 50 litres of hot water required. (1 sq. ft. per gallon). Using the heating engineers recommended standards therefore, we would build 1.4 – 1.8 sq. metres of solar absorber for each person requiring hot water. (15½ – 20 sq. ft./person). This would be equivalent to about two collectors of the dimensions described in this chapter. Thus, a three person household would require about six of these collectors, a four person family would need eight, and so on.

The more collectors we use of course, the more the system will cost. Up to a point however, the increase in energy collection that can be yielded by additional collectors will justify the extra expenditure. This is particularly so when using DIY panels which are comparatively cheap. This and other economic factors will be discussed in the next chapter which looks at the costs involved and compares these with benefits returned by solar systems.

# 13

# Economics of Solar Water Heating

The installation of a solar heating system has many advantages some of which can be measured in financial terms, and others which cannot. Houseowners who install a system will have reduced fuel bills, move a step closer to self-sufficiency, increase the value of their house and set a valuable example to their community. The community itself will benefit because each solar system installed means that less fossil fuel will be burned and the air will be cleaner and the environment more healthy. The whole country benefits as each new solar system represents a saving in fuel imports and hence alleviates economic problems caused by trade deficits. Examining effects on an even wider level, the whole world can be seen to share the rewards. The development of solar technology will help conserve the precious fuel reserves still remaining, thereby extending their benefits to future generations and easing the tensions which will undoubtedly arise between nations if there is a sudden and drastic cut in the accustomed levels of energy supply.

The advantages which are perhaps the most important are also the most difficult to quantify. How do you fix a price tag on clean air? How many kilo-watt hours of solar energy are required to prevent the outbreak of a resource war between energy hungry countries? Unfortunately we live in a society which increasingly tends to base its decisions upon quantified data and arguments based on the idea of quality tend to be ignored.

The big hold-up in the development of solar technology arises from the fact that although the sun's energy is free, the equipment for utilising it certainly is not. Thus, although the physical principles of solar collection have been understood for many years, and the equipment has been available, it has never been possible to present a convincing case for its economic viability except in specialized applications or in very sunny countries. In 1973 however, things began to change.

The oil embargo imposed by the Oil Producing and Export Countries (OPEC), and the energy crisis which it caused in the western industrialised world, brought home, in a way that no amount of argument could, the devastating effect that energy shortages can have. The subsequent five-fold increase in the price of oil has also made it very clear that oil is a finite resource which will become more and more expensive as we approach the end of its supply.

In the past few years therefore it has been much easier to convince people of the 'hidden' benefits of using a constantly renewed energy source, and with the move towards more realistic pricing of fossil fuels, it is becoming possible to show that solar heating systems can make sense even in straight forward economic analyses.

The rise in price of conventional fuels is an important point when examining the viability of solar systems. In financial terms, what you do when you install a solar system is to make a capital investment (the cost of the solar system) in order to reduce recurrent expenditure (fuel bills) in the future. The return on your investment will therefore be the savings on your fuel bill. A solar system which provides half of the energy usually consumed for heating water, should half the annual water heating bill. The more the price of conventional fuels rise, the bigger the bills become and therefore the greater will be the economic value of the returns from your solar heating investment.

In the remainder of this chapter we will put some figures to both the cost of solar system and the savings it can provide.

## Potential Savings

In the south of England, every square metre of unshaded horizontal surface receives about 1000 kWh from the sun each year. This is equivalent to the heat that would be emitted by 1000 single bar electric fires burning for one hour.

If the surface were tilted towards the south, it would intercept even more energy, say 1200 kWh/sq.m. Solar collectors are not 100% efficient of course and they cannot therefore capture all the energy which falls upon them. Taking a typical overall efficiency of 35%, we might expect each square metre of collector to yield something like 400 kWh. Increasing the size of the collector does not yield proportionately greater amounts of useful energy though. There is a law of diminishing returns because on very dull days, no amount of collectors can meet the energy requirements, whereas on very sunny days, even a fairly small collector will collect more energy than can be used. Thus 4 sq.metres (40sq.ft.) of collector area would provide not 1600 kWh, but something like 1400 kWh. of useful heat, whilst 6sq.m. (60sq.ft.) would yield some 2000 kWh.

## Value of Savings

The value of the 1400—2000 kWh. yielded by a solar system depends upon which fuel it is replacing. If you normally heat water with electricity, each kWh. the sun contributes is worth 2.6 pence at peak rates. If you use gas; the energy is much

122

cheaper. Assuming that the boiler operates at an average efficiency of 50%, each kWh. from the solar system saves you 1.2 pence.

By multiplication then, we can see that 4—6sq.m. of solar collector will save £36— £52 worth of electricity, or £17—£24 worth of gas. It is important to realise though that the actual savings will be lower if the delivered heat is not used. This will apply for instance if you take long summer vacations away from home.


## Limits for Cost Effectiveness

The simplest way of determining whether or not a solar system would be an economically worthwhile investment, is to check that the total return exceeds its total cost. The total return is a summation of the fuel bill savings for every year that the system functions. It is impossible to know in advance how long the system will function but in the calculations which follow, we will assume a working life of twenty-one years. Thus,

### Total Return = 21 x Annual Saving

Using this formula we can set an upper limit to the cost of the solar system. Systems costing more than this estimated figure, must then be judged to be investments, which in conventional financial terms are of doubtful worth. The table below indicates limits for four different systems.

| Auxilliary Fuel | 4sq.m. (40sq.ft.) | 6sq.m.(60sq.ft.) |
| --- | --- | --- |
| Electricity | £756.00 | £1092.00 |
| Gas | £357.00 | £504.00 |

*Table 13.1 Cost limits for cost-effective solar systems (assuming a 21 year life)*


## Total Cost

When making a very detailed study of economic feasibility, there are several factors which might be added to the initial expenditure in order to arrive at the total cost. An annual allowance could be made for example for maintenance.

Another less obvious aspect is the value of the invested capital itself. By this I mean that the money spent on a solar installation could alternatively have been invested and earned interest. This foregone interest should then be added to the total cost. If on the other hand money was borrowed to pay for the system, the loan charges should be added.

If the capital cost was borrowed on mortage, account must also be taken of the fact that the buyer will benefit from the increased tax allowances which go with increased mortgage payments.

The taxation system offers a further incentive (unintentional of course) for investment in solar systems. The returns on most alternative investments will be classed as taxable income. The returns from a solar installation come in the form of heat and reduced bills and thus cannot be taxed.

## Future Savings

The accurate evaluation of future savings from a solar system is not easy because of the difficulties inherent in predicting the future. A great variety of financial jargon will be encountered in any of the detailed analyses which attempt this, but it is worth remembering that no final and definite answers can result. What we can get is a mixture of calculated guess work plus a touch of crystal ball gazing which will indicate some possible results.

'Discounted cash flow analysis' is the off-putting name given to a widely accepted method of reducing the value of future savings in order to determine their equivalent present-day worth. This discounting assumes that money can be profitably invested. For example with a 10% interest rate, £10 today can be worth £11 in a year's time. The converse of this, that a saving of £11 next year is worth only £10 in present-day terms is the principle behind discounted cash flow analysis.

Economists however, have often chosen unrealistically high discount rates. They have assumed that invested money can earn high returns. In actual fact, during the post-war period, the real returns on investment, after inflation has been accounted for, have been low and sometimes even negative. In such circumstances it is misleading to reduce the value of future savings by any great amount. Certainly the discount rate of 10% used by government departments seems ridiculously high.

## Inflation

Whereas high interest rates urge against long term investments, high inflation rates reinforce the argument in their favour. For example, when the rate of inflation exceeds the bank interest rate, money deposited in the bank will actually lose value, in spite of gaining interest each year. In such cases investment in solar systems forms a more attractive alternative as the returns on the initial capital will keep abreast of inflation. As fuel bills inflate, along with all other goods and services, so also will the value of the solar contribution.

A further point in favour of solar investment is that the price of energy has been rising even faster than general inflation and is likely to do so in the future. Thus the returns from a solar system will not only increase in nominal terms due to general inflation, but also in real terms due to the increase in the value of energy.

## Pay-back Period

In discussing the economic viability of solar systems, one often hears mention of 'pay-back periods'. This is simply the number of years for which the system must operate before the value of the fuel savings which it accrues equals its cost.

124

Using a simple method of analysis, taking no account of maintenance and assuming that benefits of inflation are balanced by disadvantages of foregone interest, we can say that a system costing £250 and contributing a saving of £25 a year, will have a pay-back period of 10 years. A system costing £500 and still contributing the same £25/year, will take 20 years to pay back its initial cost.

**Pay-back Period (years) = Total Cost ÷ Annual Savings**

The concept of pay-back periods emphasises the need to install systems to such a standard that they will have a long life, at least long enough to repay the investment made in them.

## Commercial Systems

You can buy a solar water heating system with about 4sq.m. (40sq.ft.) of collector area, and have it installed for between £500 and £700. With reference to the table we have compiled (table 13.1), it can be seen that such prices will make the investment profitable only if the system is augmenting an electric water heating system. Commercial systems are still too expensive to compete with gas.

It should be understood that where a gas supply is available, the fitting of a small gas circulator to replace an electric immersion heater is a much more cost-effective measure than installing solar collectors.

## Future Improvements

Several things could happen which would bring about an improvement in the economics of commercial solar systems. The most significant would be through political intervention. All governments now realise that saving energy is in the national interest. Several, such as those in the USA and France, are beginning to encourage energy conservation by providing tax relief to those who invest in energy saving improvements to their buildings. In other countries such encouragement is more indirect. The Dutch government has recently announced that the price of natural gas is to be raised, not due to increased production costs, but in order to conserve supplies.

In the UK, the government's Select Committee on Science & Technology has published a recommendation that grants be provided to subsidise houseowners wishing to install solar water heating systems. This could be done under similar legislation to that which provides grants for general house improvements. This is a subject currently under discussion and mass lobbying of local Members of Parliament, local authorities and the Department of Energy could have a positive effect.

Such legislation would rapidly increase the market for solar collectors and we would hopefully begin to see the drop in prices which one expects when mass production facilities come into use.

Finally, the flourishing solar industry which could then evolve would be a fertile ground for the development of cost-effective high efficiency collectors of the types described in Chapter 4.

## Economics of DIY

Having determined that considerable changes are necessary before commercial systems become cost-effective for the majority of homes, what can be done here and now? The answer is to employ yourself on the job.

You could buy solar panels and the necessary accessories and then install them yourself. On-site labour charges usually account for about 1/3 of the cost. You could even go the whole hog and construct the collector panels yourself. This would result in much greater savings. Another alternative would be to buy the absorber plate ready-made and fix it directly between the roof rafters under patent glazing. The absorbers plates usually cost about £30—£35/sq.m., around half the price of the cased and glazed panels.

## Golden Rule

Before going any further with details of the savings which are possible through DIY, there is one golden rule you must adopt if you are to reap the economic benefits of self-help; YOU MUST ENJOY YOURSELF. If you are substituting your own time and labour for that of someone else, it might be argued that you should add to the initial cost, a sum based on how you value your own time. So you may end up with a total cost almost as high as a commercial installed system. Unless that is, you happen to enjoy the work. If you think you might find climbing on roofs, crawling through attics, bending pipes, tightening nuts, satisfying activities, then self-help solar heating becomes your hobby and no one should suggest that the system you build costs any more than the material you bought.

## DIY Installation costs

Installation charges are usually about £150—£200. The DIY householder however will have to buy accessories such as pipe, pump and fittings at a higher price than the professional installer due to buying in small quantities, not knowing the cheapest local source of materials and perhaps through missing trade discounts (particularly if materials are bought in DIY supermarkets as opposed to builder's merchants). Taking these factors into account, the savings to be made through DIY installation are more likely to be in the order £75—£125. As an example, a system I installed in a London terrace house cost £469 in materials alone. This still exceeds the gas-derived limits for cost effectiveness as indicated in table 13.1 It was a two tank, pressurized system. All the pipework was copper as were the commercially manu-factured solar absorbers covering an area of 3.6 sq.m. The cost breakdown was as follows:

    3.6 m$^2$ Solar Collectors (glazed)....................... £260
    200 litre Cylinder (indirect)............................ £56
    Copper Piping....................................... £28
    Fittings........................................... £28
    Pump & Isolating Valves.............................. £25
    Anti-freeze........................................ £7
    Pressurized Vessel & Filling Assembly .................. £35
    Pump Control....................................... £30

126

## DIY Collector Construction Costs

The type of collectors described in chapter 12 could be made in a home workshop for between £16.50 and £33.00/sq.m. The casing costs about £14/sq.m. (£1.40/sq.ft.) and a self built copper absorber costs some £19/sq.m. (£1.90/sq.ft.) This is the same as the cost of a new central heating radiator, but much more than that of a second hand radiator which can be obtained for about £1/sq.m. but requires cleaning and fittings which increase its cost to about £2.50/sq.m. (£0.25/sq.ft.)

Assuming the same prices for ancilliary equipment as those listed above (quite an expensive system), the total cost of a system with 4 sq.m. (40 sq.ft.) of self-built collectors would be £275—£341. If you refer back to the table showing cost limits for cost effective systems, you will see that the upper limit for a 4 sq.m. system replacing gas is £357. The home-made-collector-systems would all meet this requirement. The pay-back periods are however rather long — 16 or 20 years when derived from current gas prices. The problem lies in the cost of ancilliary equipment. This takes the lion's share of the total cost in a DIY system. Its proportion can be decreased by increasing the area of collectors used. This gives a higher energy yield without significantly increasing the amount of relatively expensive ancilliary equipment.

We find therefore that if we use 6 sq.m. (60 sq.ft.) of DIY collectors the system will cost either £308 or £407. Both figures are well within the upper cost limit of £504 for 6 sq.m. systems replacing gas and their pay-back periods are three years shorter than those of the 4 sq.m. systems.

## Less Sunny Regions

There are few countries which have less sunshine than the UK and those which do,

| Collector Type | | Initial Cost | | |
| --- | --- | --- | --- | --- |
| | | 4sq.m. (40sq.ft.) | 6sq.m. (60sq.ft.) | 6sq.m. Double Glazed |
| Self-Installed Commercial Panels | | £469 | £534 | £594 |
| DIY 2nd Hand Radiators | | £275 | £308 | £338 |
| DIY Copper Tube & Sheet | | £341 | £407 | £437 |
| Max. Cost Effective Investment (GAS Price DERIVED) | Scotland | £294 | £409 | £451 |
| | S.England | £357 | £504 | £554 |

*Table 13.2 DIY Solar System Costs Compared with Maximum Cost Effective Investment Recommendations.*

127

like Greenland and Lapland, are associated with wintry conditions and are generally sparsely populated. For solar systems to achieve economic competitiveness in Britain then, suggests that they will be competitive just about everywhere.

You may remember though that the calculations which have been carried out in this chapter were based upon conditions in the south of England where we might expect to receive some 1000 kWh/sq.m./year. The map in appendix 2 (figure A.9) indicates the variation which occurs throughout the country. In Scotland for example there is a decrease of almost 20% in the annual total of energy received. This means that there will be a corresponding decrease in the savings from the system. We must therefore reduce the upper limits for cost effectiveness when considering more northerly applications.

In table 13.2 these adjusted limits are shown underneath the estimated costs of various types and sizes of solar systems. It is reassuring to see that the home-made-collector-systems still seem to be economically worthwhile north of the border so long as they are large enough, say 6 sq.m. in area. The additional cost of double glazing the collectors has also been estimated. Assuming that the improvement in performance is 10% the cost/benefit is so marginal that it does not seem to be worth the bother of adding an extra sheet of glass.

### Fuel Price Increases

All the calculations made so far have been based upon the assumption that fuel prices will not rise in relation to other goods and commodities, i.e. that they will merely keep abreast of inflation. If in fact fuel prices do rise, this will strengthen the economic case for investment in solar systems. The more fuel prices increase, the quicker will solar systems be able to pay back their initial cost. This is shown in table 13.3 for gas prices, and in table 13.4 for electricity prices.

| Collector Type | Zero Annual Increase | | 4½% Annual Increase | | 7% Annual Increase | |
|---|---|---|---|---|---|---|
| | South England | Scotland | South England | Scotland | South England | Scotland |
| 6 sq.m. DIY 2nd Hand Radiator | 13yrs. | 16yrs. | 10yrs. | 12yrs. | 9yrs. | 11yrs. |
| 6 sq.m. DIY Copper Tube & Sheet | 17yrs. | 21yrs. | 13yrs. | 15yrs. | 12yrs. | 13yrs. |

Table 13.3 Pay-back Periods of DIY Solar Systems for Different Rates of Increase in Gas Price.

| Collector Type | Zero Annual Increase | | 4½% Annual Increase | | 7% Annual Increase | |
|---|---|---|---|---|---|---|
| | South England | Scotland | South England | Scotland | South England | Scotland |
| 6 sq.m. DIY 2nd Hand Radiator | 6yrs. | 7yrs. | 5yrs. | 6yrs. | 5yrs. | 6yrs. |
| 6 sq.m. DIY Copper Tube & Sheet | 8yrs. | 10yrs. | 7yrs. | 8yrs. | 6yrs. | 8yrs. |

*Table 13.4 Pay-back Periods of DIY Solar Systems for Different Rates of Increase in Electricity Prices.*

**Other Energy-Saving Investments**

Other energy-saving measures will share the advantages such as inflation-proof, untaxable returns which have been outlined above for solar systems. Indeed many provision will prove to be considerably more cost-effective than a solar water heating system. Draught-stripping, loft-insulation, additional cylinder insulation, pipe-lagging and foam-filling of cavity walls all offer greater returns on the cash spent.

**Conclusions**

It is worth remembering that there are many benefits which arise from using solar energy which cannot be quantified and included in a financial appraisal. Even taking those factors which we are familiar with evaluating in financial terms, difficulties arise due to the uncertainties surrounding a system which will be operating for many years into the future.

Nonetheless, making the assumptions which have been outlined in this chapter, we have arrived at two important conclusions.

1) Commercial solar systems for heating domestic water have questionable financial value, except where electricity is the only alternative fuel. It has however, been indicated how this situation might change within a very short time.
2) Self-installed systems using DIY solar collectors appear to be financially competitive with gas even at its current price. This is true not only with salvaged old radiator panels, but also with all-copper absorbers and central heating radiators bought at their retail price. The financial analysis has demonstrated that it pays in the long run to use large areas, say 6 sq.m., when using DIY or other cheap collectors.

Finally it is important to realise that any solar heating installation is essentially a long term investment. When replacing electricity as a water heating fuel, a DIY system will require between 5 and 10 years to pay back its initial cost. When replacing gas, which is of course much cheaper, it may take twice as long. The more fuel prices rise of course, the shorter these pay-back periods will become.

# 14

# Swimming Pool Heaters

Solar energy by its nature is dispersed over a wide area. Energy consumers, or people, are likewise spread over large areas. The economics of scale therefore, which have been the excuse for all kinds of social and environmental disruption, do not apply in the case of solar energy exploitation. That is to say there is no great advantage to be gained from having huge solar energy collection centres, which will only have to redistribute the energy to individual homes.

This is in stark contrast to our conventional energy supply systems which are characteristically large in scale, highly centralised and can only be controlled by similarly large scale bureaucracies. Their structure makes them vulnerable both to attacks from minority groups seeking disruption and to manipulation by autocratic figures in positions of authority. Thus energy supply, the lifeblood of technological society can be seen not only as a question of engineering and resource management, but also as a weapon to be wielded in political and social struggles.

With such thoughts in mind, many recent supporters of solar technology development have seen the move away from total reliance on the national power grid, not only as a symbol of breaking away from an unjust society ruled by distant authority, but as a positive and enabling step towards the establishment of a decentralised, cooperative utopia.

Given such aspirations and hopes, it may seem somewhat ironic that the first practical applications of solar heating have been in the heating of swimming pools for luxury hotels.

In engineering terms though this is hardly surprising. For six months of the year, swimming pool water does not have to be heated any more than 10°C (18°F) above the air temperature. This means that swimming pool solar collectors can operate at low temperatures, losing very little of the heat they absorb to their surroundings, and thereby achieving high efficiencies. Swimming pool collectors raising the water temperature by only 1 or 2°C at each pass, usually operate at efficiencies in the range 70-80% compared with the typical 30-40% achieved by domestic water heaters. Not only do they yield more energy, but swimming pool collectors can cost much less than those designed to operate at higher temperatures. This is because the solar absorption plate, being only a few degrees above the air temperature, has such a low level of heat loss, it does not benefit significantly from layers of insulation or glazing. Indeed a transparent cover, which will reflect at least 16% of the solar radiation may actually reduce energy collection, except in cases where the absorber is exposed to frequent strong winds.

The absence of such a protective cover also enables the collector to operate on some occasions when there is no direct solar heat gain. This is possible because heat can be absorbed from the air which will sometimes be warmer than the water. System costs are further reduced by the fact that only the collection and circulation elements are additional, the pool itself fulfilling the function of heat storage. Finally swimming pools have the advantage that, unlike most energy demands, the demand follows closely the supply of solar energy. That is, the sunniest months are those in which swimming pools are most in use.

## VALUE OF SOLAR HEATING POOLS

The value of a solar heating system will of course vary not only with the size and location of the pool, but also with pattern of use to which it is put, and with the users' personal preferences. Some people for example especially

value the bracing effect of a cold dip and you would have a hard time convincing them that any expenditure on heating is worthwhile.

To generalise one can say that there are three categories of pools:
 i) Unheated pools
 ii) Heated Pools
iii) Pools closing during summer

### Unheated Pools

Unheated pools are generally used only for about three to four months in the year. A solar system connected to such a pool would lengthen the swimming season by about two months. When compared with a fossil fuelled heating system, it would have the disadvantage that it would be unable to reheat the pool rapidly after dull periods and initial warming up in the spring would take several weeks. In addition the initial cost of a commercially installed solar system would perhaps be double that of a conventional boiler. The price of fuel however would whittle away the boiler's financial advantage in a short number of years. If the pool is to be used only during the summer months, there is no doubt that a solar installation would be the best choice. If the pool is to be used throughout the year, a conventional boiler will be needed to achieve comfortable bathing temperatures and consideration should then be given to the points outlined in the following paragraph.

### Heated Pools

Heated pools use large amounts of

expensive fuel the price of which will continue to rise in the years ahead. In just about any form of analysis comparing the cost of a properly designed and installed swimming pool solar heating system with the savings it provides in the form of reduced fuel consumption, the system proves to be value for money. This is true both for the case where one has in hand the money to pay for the installation outright — the fuel savings outweigh the cash return which could have been realised through investing the money in a Building Society — and the case where the price of the installation has to be borrowed at 12½% interest rate.

**Pools Closing During Summer**

This third category of pools is the only one where solar heaters may not be a worthwhile investment. There are still many schools where the swimming pools are unused during the summer vacation. This is of course a ridiculous waste of a valuable amenity which has probably been paid for in part or whole by public funds. Nevertheless it does happen and until education authorities and local clubs can cooperate to improve the situation, a careful study of the economic factors involved is required before committing large sums of money for the provision of solar heating. The problem is of course that the bulk of the savings that a solar system provides is contributed during the very months that schools are closed for holidays.

**INSULATING COVERS**

In all categories of swimming pools the

first step in heating the water and reducing costs is to provide a pool cover. Once the pool has been initially heated at the beginning of the season, the main heating requirement is that of replacing heat lost through the pool surface. Heat is lost by radiation to the night sky and by convective currents and the evaporation of pool water. Compared with these factors, heat loss into the ground, in a sunken pool is insignificant. The cheapest form of cover is probably a layer of black PVC. 'Airwrap', originally manufactured as a packaging material is an attractive alternative. It is a double skin of clear polythene with a matrix of air bubbles trapped between them. This has the advantage that it will not sink in the water and it allows solar radiation to pass through it. Pools which are more than a metre deep are reasonably good absorbers of solar radiation in their own right. A translucent plastic therefore will act as an insulative cover and still allow the pool to heat itself to a limited extent on sunny days. A family I once visited told me that they had tried various commercial covers in the past and then showed me the home-made version which they found preferable. They had stretched polythene (a U.V. resistant grade would be best) across three rectangular timber frames which spanned the breadth of the pool. This left an air space of about 300 mm (1ft.) above the water surface and hence it was possible to go swimming with the cover still on in rainy weather.

**COLLECTOR SIZE**

Most manufacturers recommend using an area of solar collectors equal to half the pool's surface area. It may be worth expanding this area if the collectors have to be mounted more than 25⁰ away from south or if they are mounted at a very steep angle. The temperature in a swimming pool heated with collector areas equal to and equal to half of the pool surface area are shown on the accompanying graph and can be compared with the temperatures in an unheated pool which are also indicated.

## COLLECTOR ORIENTATION

The ideal orientation is likely to be slightly west of south although the orientation of unglazed absorbers is even less critical than conventional panels which suffer from the increasing reflection at the glass surface when the radiation strikes at progressively more oblique angles. For the best results stay within 25° of south.

## COLLECTOR TILT

Generally a shallow tilt is desirable as the system will operate in summer when the sun is high in the sky. A steeper angle might be desirable in sheltered pools which may achieve quite satisfactory temperatures without much help during mid-summer, and also in combined systems (see end of this chapter) where the collectors are serving another energy demand in addition to the pool. All angles between horizontal and 60° are acceptable. They can even be mounted vertically although in such cases the area should be increased by over 50%. Remember also that the further away the collectors face from south, the shallower should be their tilt angle.

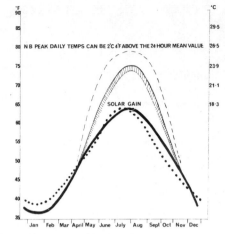

**··· AIR TEMPERATURE**
**— WATER TEMPERATURE**
(Unheated pool)
ⅢⅢⅢ WATER TEMPERATURE WITH SOLAR PANELS HAVING COLLECTOR AREA = ½ AREA OF POOL.
– – – WATER TEMPERATURE WITH SOLAR PANELS HAVING COLLECTOR AREA = AREA OF POOL.

*14.1 Indicative pool temperatures for an unheated swimming pool and a solar heated pool with panel area equal to half of the pool area and a solar heated pool with a larger area of solar panels equal to the entire pool surface area.*
*N.B. This graph was prepared some years ago by Robinsons Developments. Their recent experience suggests that temperatures are in fact likely to be higher than those indicated.*

133

## CONNECTING THE SOLAR SYSTEM

There are many variations in the way a solar system can be connected to a swimming pool. The illustration indicates a simple method which involves breaking into the existing filtration circuit and thereby saving the expense of an additional pump. The existing pump may of course not be suitable to take the additional load of circulating through the collectors but this is unlikely as their flow resistance is small in comparison with that of the filter unit.

Following the circuit indicated, water is drawn off the pool from sump (1) and or the skimmer (2) by the pump (3). It is then passed through the filtration unit (4) and from there, it would normally be returned to the opposite end of the pool (5). When the solar system is added however, the water is first led to a three way, motorised valve (6). This is electrically controlled (8) and (9) so that when there is energy available in the solar panels, the stream is diverted to their bottom inlet (7). When there is no heat to be gained from the collectors, water is directed to the pool as before.

The collectors are connected in parallel and heated water is drawn off at the top corner (10) diagonally opposite the inlet. A drain-cock should be fitted in the remaining bottom corner (12) which should be the lowest point on the collector bank, and it is usually advisable to have an air vent at the highest point (13). Solar heated water is returned to the fil-

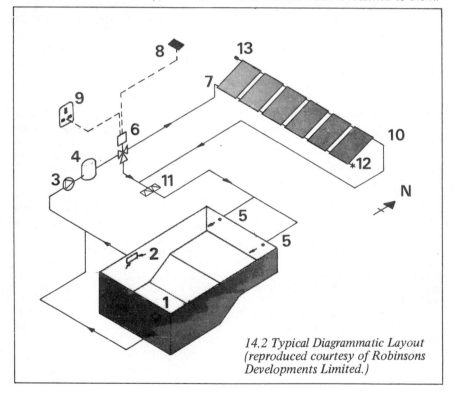

*14.2 Typical Diagrammatic Layout (reproduced courtesy of Robinsons Developments Limited.)*

tration circuit at a point before any additional heating unit (11).

It may be possible to economise on the capital cost of such a system by using two solenoid valves fitted in reverse in place of the expensive motorised valve. Running costs could be lowered by installing a separate pump for the solar circuit. This could be of a much lower power rating than the filtration circuit pump which would then be used only during the few hours each day which would suffice for filtration.

## SWIMMING POOL COLLECTORS

The major differences between pool collectors and standard flat plate solar collectors are their lower operating temperature and the increased danger of corrosion.

As already stated, for low temperature applications such as pools, uninsulated and unglazed absorbers are usually adequate. This of course means that the exterior of the abosrber plate is exposed to weathering. Furthermore, the swimming pool systems are almost invariably direct circuits. (At the low operating temperatures which are necessary to maintain maximum efficiency, a very large and expensive heat exchanger would be necessary in an indirect circuit). This means that not only will the water passing through the collectors be aerated, but it will also contain the corrosive chemical compounds which are formed when small quantities of urine in the water interact with the pool's chlorination.

Given these added dangers extra care must be exercised in the choice of materials. Copper and certain grades of stainless steel would probably be the most resistant to corrosion. They are however expensive. Aluminium and steel are liable to corrosion under such hostile conditions and we are therefore left with plastic collectors appearing as the best choice. The major problems encountered with plastics will be degradation due to ultra-violet radiation, softening and deformation at high temperatures and leaching of their stabilisers by hot water. Given these problems the best choice of plastics would be polypropylene and polycarbonate pigmented with carbon black, not only to increase absorption, but to resist the effect of U.V. radiation. A recent new design of swimming pool collector incorporates a plastic pipe coiled in a slab which can be laid as paving for the pool surrounds.

## DIY POOL COLLECTORS

Attempts have been made to make polypropylene sandwich type absorbers from Corex, a semi rigid wafer manufactured originally for packaging. The difficulty however comes in making a water tight bond between the open end of the wafer, and a header pipe. I have been told that this can be done with Dow Corning silicone sealant as long as pressure on the joint is low. I have not however seen such a joint tested.

A more proven DIY approach would be to use corrugated aluminium sheets, with an annodised or stoved on coating for corrosion prevention, and construct a trickle type collector as described in Ch. 4 and Ch. 15. Due to the high evaporation losses which would occur with an open system such as this, it would be advisable to enclose the abosrber with a sealed transparent cover.

A closed version of the trickle collector which would not need a transparent cover has been described by Dr. Cleland McVeigh in his book "Sun Power". It uses 'Airwrap', the polythene/air bubble packaging material, mounted on a ply wood sheet in place of the corrugated metal. The air bubble matrix distributes the flow of water which is released by a header pipe along the upper edge. A sheet of black butyl is then spread across the assembly and sealed at the edges except along the bottom where the heated water is collected.

Finally it is worth mentioning that the solar absorber can be an object of beauty. It is possible to raise the pool temperature by cascading water over a series of black painted concrete terraces at the pool side.

### PUMP CONTROL

Electronic differential temperature controllers as described in Ch.5 can be used but for swimming pool applications they must be more sensitive than usual due to the very low temperature differences at which it is still worth circulating. Greater sensitivity of course means higher cost.

Some control is necessary however as a system left running 24 hours a day would actually have a net effect of cooling a pool due to radiation to the night sky. The cheapest automatic control would be a time switch which turns the system on in the morning and off at night.

Another alternative which would be

cheaper than a differential controller, is a device designed and manufactured by Robinsons Developments, who also make polypropylene absorbers. This switches on the pump when the sol-air temperature reaches a pre-determined level. 'Sol-air' indicates a composite reading of radiation, humidity, wind, and air temperature. The device itself is quite simple though, consisting of a thermostat encased in a black metal box and placed in front of a polished metal reflector.

### COMBINED SYSTEMS

A large bank of solar collectors serving a summer load in the form of a swimming pool, may be able to make a useful contribution to a space heating load during the other half of the year. Similarly, a domestic water heating installation will often collect more energy in summer than it needs and such surplus could be diverted to a swimming pool.

The electronic controls for such diversion of the solar heat supply are available from Don Engineering Ltd. and where pools are sited close to buildings, such combined systems would make sense. One would have to rethink the questions of glazing, insulation and tilt angles, bearing in mind which system is considered to have most priority.

#### Heat Pumps

Much work is being carried out on the development of heat pumps for space heating. Heat pumps are devices which extract heat from a source which is at a low temperature, and upgrade this heat to deliver it at higher temperatures. Refrigerators do this. They extract heat from their interiors, which start out at room temperature, and expel the heat in the form of warm air at the rear.

To carry out this operation, the heat pump requires an energy input, usually electrical. To be worthwhile of course, the device must 'pump out' more energy than it consumes. Typically they can deliver about 3kWh for every 1 kWh of electricity they use. The proportion will become less favourable however as the temperature difference between the heat source, and the delivered heat increases.

## Pools & Heat Pumps

This is where swimming pools may prove useful. Normally the heat source is the open air, or a stream, or the ground itself. A swimming pool, particularly one connected to a solar system, is likely to be at a higher temperature than the air or a stream and will be more conveniently tapped than the ground. The heat pump will also complement the performance of the solar collectors. By reducing the pool's winter temperature, it will lower the collector's operating temperature therby increasing its efficiency and extending its period of useful functioning.

If a new pool is being constructed, it is worth siting it near a building and installing ducts which could be used during the next decade for connecting a heat pump when the technology is developed.

## Solar Ponds

It has been mentioned that the pool itself can collect solar energy. Much of the absorbed heat is however raised to the surface by convection currents and there lost to the air. The efficiency of an energy collecting pool can be greatly increased if the convection currents can be suppressed. Achieving this is to dissolve salt in the water in such a way as to form a gradient of salinity, with a saturated solution at the pool bottom and much less salty at the surface. In such a pond, absorbed heat would remain at the bottom and high temperatures could build up.

A theoretical study has suggested that a solar pond in London with a surface area equal to the floor area of a house, could heat that house throughout winter at a cost comparable with that of gas if the installation was paid for on a twenty year mortgage. The pond in the study was three metres (10ft) deep, an upper layer of salt saturated water two metres deep, and underneath a separating membrane one metre of plain water.

Will swimming pools in the future no longer be a recreational luxury but a necessary part of keeping warm in a fuel-deprived world? Shall we see new houses being built in groups around artificial ponds which not only keep the children happy during summer, but also soak up the summer sunshine and waste heat from the buildings to return it during cold spells? These questions cannot be answered now but they may be answered sooner than you think. Which would you opt for if given the choice between a backyard solar pond and a central heating boiler?

# 15

# Examples

## Introduction

By now you should have a clear idea of how a solar system works what its components are and the sort of work involved in installing one. You may however still have doubts about whether they really do work, and whether you could build one yourself.

Everyone I know who has built a system has had a real kick from the first trickle of warm water which proves the theory. The first solar shower of course is ecstatic.

This chapter introduces some of these systems. Some have been built by people with little experience, and others by people with no previous experience, but learning fast with a little help from their friends.

In the preceding chapters I have tried to keep to describing only techniques which I, or close associates, have personally tried. This has meant that there has been only brief references to some inter-

esting alternatives such as trickle collectors and air heating collectors. For this reason the features on the two systems which use these collectors, which have advantages of economy and simplicity, are of special interest.

Perhaps the greatest value of these examples though will be the opportunity to hear of the mistakes others have made. It has been said that experience is the name we give to our past mistakes. After you have read about some of the goofs I have made, you will understand why I felt 'experienced' enough to prepare this book.

## Evening Class Installation

In 1977 I ran a series of evening classes in Kingston for people interested in applying solar energy. In order to produce at least one working example during the limited time available, it was agreed at the beginning that one of the participant's houses should be chosen as a pilot scheme with which everyone would assist.

15.1 Solar collectors installed by evening class participants in Kingston.

The selected house was one with a projecting ledge at first floor level on the south side, which would allow the collectors to be mounted below the level of tanks fitted in the attic. The ledge also had easy access from the garden and the first floor bedroom window. Hot water in the house was normally provided by a coal fired boiler in winter and by an electric immersion heater in summer — electricity is of course the most expensive fuel for heating water. Hot water requirements were not so great though, as the family's children had grown up and left home. Thus a smaller than normal collector would suffice.

The solar panels, each having an absorption area of just over a half square metre (6 sq. ft.) were made up in the class workshop by three separate groups working to the same design. They are second hand radiators encased in a box made up from old floor boards and a backing sheet of marine ply (exterior quality ply would have been adequate and cheaper). A single sheet of window glass formed the cover which was secured by aluminium angle screwed to the side of the case and sealed with silicone sealant applied from a gun. The decision as to when to fix the glass is a difficult one. Ideally it should not be done until the whole system has been installed and tested. Unexpected rainfall however puts paid to many ideal plans, as those of us who sat through a hail storm, shivering, and holding sheets of polythene over the unglazed collectors to protect the insulation quilt from a soaking, will remember. As it happened, we still had to remove the glass from one panel several weeks later when a leak from one of the radiator connections was noticed. Unfortunately some of the sealant had seeped through from the glass/aluminium joint, to the glass/timber frame joint making removal very difficult. In fact, I cracked the glass in the attempt.

In the attic, a 150 litre (30 gallons) indirect cylinder was installed with two plastic tanks supported on a platform fixed to the rafter ties above it, serving as cold feed and expansion vessels. The hot outlet from the solar cylinder was connected to the pipe feeding cold water from the existing storage tank to the existing hot water cylinder in the bathroom cupboard.

The size of the solar collector was limited by the dimensions of the projecting ledge — in fact as can be seen in the photograph, one of the panels had to be mounted on its side due to my failing to allow enough space between the panels for pipe bends when measuring up the site. A collector which is small in relation to its storage tank never produces very dramatic temperature rises, it functions purely as a pre-heater.

"We never get really hot water from the solar system alone," say Joy and Peter who live in the house, "but we have noticed that it takes a lot less time for the cylinder in the bathroom to heat up when we want a bath."

Although the panels were largely completed during four evening classes, it required a further six Saturday afternoons to complete the mounting and plumbing-in. Several memorable mistakes were made which are worth recounting.

The primary circuit, that linking the collectors to the heat exchanger in the solar cylinder, was the first to be completed. Curious to test joints made by housewives turned plumbers, we filled the system via the expansion tank. We became even more curious when after having poured in about fifty litres (10 gallons), the circuit still had not become

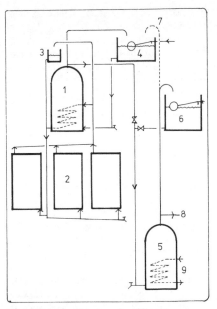

*15.2 System diagram for Kingston evening class installation.*
1) *Solar cylinder*
2) *Solar collectors*
3) *Solar circuit expansion tank*
4) *Cold feed tank*
5) *Existing hot water cylinder*
6) *Existing cold water storage tank*
7) *Extended vent pipe from existing hot water cylinder*
8) *Pipe leading hot water to taps*
9) *Connections to coal-fired boiler*

full. When somebody noticed that the cylinder was becoming heavier, we realised where the water was going. After cursing the manufacturer for selling us a cylinder with a leaking heater exchanger, I looked over the pipework preparing to disconnect it, and discovered that the fault was actually my own.

The inlet and outlet for heat exchangers are usually on the same side of the cylinder. In this one however, they were on opposite sides. I had misdirected someone to connect the return pipe, going to the collector inlet, to the cylinder connection intended for the secondary circuit's cold feed.

When you are connecting an indirect cylinder, remember that the connections to the exchanger coil are normally external, or male, threaded whilst connections into the tank itself are internal, or female, threaded. To be certain, just stick your finger in the hole to feel the coil.

Later, when the secondary circuit, that linking the new cold feed tank to the existing hot water cylinder via the solar cylinder, was filled, we had another mysterious reaction. We had already sat down triumphantly for a break before tidying up, when Peter noticed water was pouring out of the overflow pipe projecting through the roof. This was coming from existing cold water storage tank. At first I thought we had disturbed its cover which was resting on the arm of the ball cock and preventing it from cutting off the flow of mains water into the tank. Unfortunately it was not so simple. Eventually it was discovered that water was gushing out of the expansion pipe coming from the top of the existing cylinder in the bathroom, and discharging over the cold storage tank. This was overfilling the cold tank which then discharged through the overflow pipe on to the roof.

With the solar system connected, the old cylinder was now being fed water from the new cold feed tank above the solar cylinder. You can see this on the circuit diagram. The new cold feed tank is considerably higher than the old one so there is more pressure on the column of water in the expansion pipe rising out of the old hot water cylinder, enough in fact to make it continuously overflow. The expansion pipe had to be extended therefore to above the level of water in the new feed tank. The alteration is dotted on the circuit diagram.

In spite of the mistakes and the consequent delays, this was an enjoyable job and I am glad to have worked with the many people involved. I believe there is much potential in the framework of local authority adult education classes. They can provide technical assistance, workshop facilities, and perhaps most important, a chance for productive co-operation with other interested people in the neighbourhood. There are of course very few courses on DIY solar heating offered as yet but if the local authorities received sufficient numbers of letters requesting it more would be organized.

## Council House Retrofit

This was the first installation I carried out using commercially manufactured solar collectors. It arose from a research programme I was carrying out at the Architectural Association to investigate the potential for using solar energy in existing houses. It is an indirect, two-tank, pressurized system intended as an example of how solar water heating systems could be incorporated in the rehabilitation programmes being carried out by most local authorities to improve their housing stock.

Four Sunstor panels, an all copper absorber measuring 1500 x 600mm (5 x 2ft.) in an aluminium casing, were mounted on a tubular steel framework which spans across the existing gulley roof whose pitches faced east and west.

The fact that the collectors had to be mounted wholly above the attic meant that there was no space to locate an ordinary expansion above them. A sealed system with a pressurized vessel to take up any expansion was the only practical choice. The pressurized vessel can be fitted below the level of the collectors and hence in a protected area inside the house or attic. Pressurized systems do increase the initial cost of the system but they will be less prone to air infiltration and hence slower to corrode.

The collectors were glazed before being lifted on to the roof. With light-weight collectors this is no problem and can be advantageous if the glass is fixed in pre-formed gaskets as in this case. Squeezing the last side of the glass into the gasket can be a tricky business. It is rather like fitting a tyre on a bicycle wheel, except that the metal rim of a bicycle wheel is a lot less fragile than glass. I used glass with ground edges and I think I would have otherwise have ended up with some bright red stains fouling up the transmissivity of the glazing if I had stuck to using plain cut glass with rough edges. Glaziers will do this for about 50 pence/metre or you can smooth the edges yourself with an oil-stone.

*15.3 Solar collectors installed on a modernised London council house.*

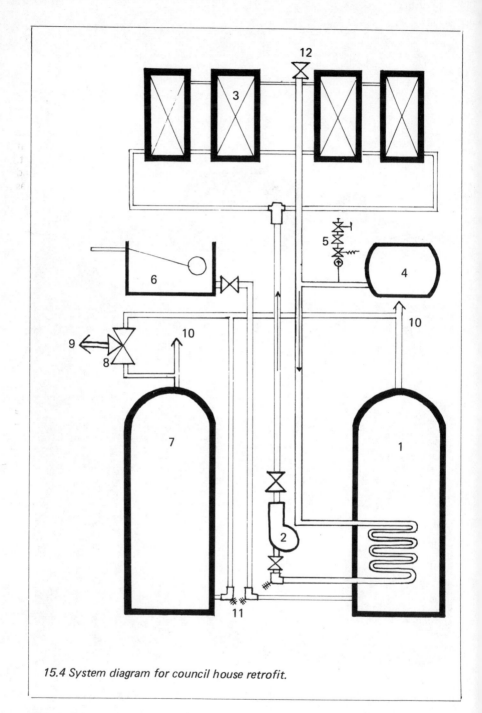

*15.4 System diagram for council house retrofit.*

**15.4 System diagram for council house retrofit**

1) Solar Cylinder
2) Circulator
3) Collectors
4) Pressurized expansion tank
5) Pressurized filling assembly
6) Cold feed tank
7) Conventional hot water cylinder
8) 3-way diverter valve
9) Pipe to taps
10) Vent pipes discharge over cold feed tank
11) Drain cocks
12) Air cock

For connecting the collectors to each other, I used soldered capilliary fittings. In future I would always choose compression fittings for this purpose. The first problem was a howling wind which made it very difficult to heat the job sufficiently to melt the solder. Secondly, when the temperature did begin to build up, the neoprene gromets, which sleeve round the connection pipes sticking out of the aluminium casing, began to melt. With molten neoprene dripping round the ednd of the capilliary fittings, it was impossible to see the appearance of the ring of silver solder which signals that the joint is complete.

Inside the house, the solar cylinder was located in an alcove between a chimney breast and the corner. This proved satisfactory but the conventional hot water cylinder which was part of a prefabricated combination unit including a cold storage tank has caused inconvenience to the residents. This was situated in a bedroom adjacent to the room in which the solar cylinder was placed. The noise from the cold tank refilling via the ball-cock — rather like the sound of a W.C. being flushed — can be disturbing if your lying in bed.

A couple of times during the installation, I was responsible for soaking the floor. Fortunately no damage was caused. The first occasion was due to a drain-cock being located in a barely accessible position. Not only was there difficulty in reaching it, but unlike most drain-cocks I had dealt with, it had a locking nut which I failed to notice until after it had jammed the cock in a slightly open position. It soon became apparent after the system was refilled! Remember that drain-cocks have to be easily accessible if you want to avoid making a mess.

The second leak came from a hair-line crack which appeared around the joint where the cold feed pipe entered the solar cylinder. Care must always be taken when tightening cylinder connections as the thin copper sheet used in the fabrication of modern tanks warps easily. In this case I had fitted a second hand tank which had obviously been warped too often. The whole system had to be drained down, fittings disconnected and the cylinder replaced with a new one costing £55.

As is normal with two tank systems, the heated water from the top the solar cylinder is lead to the bottom of the conventional hot water cylinder. If it has not been sufficiently heated by the sun, a thermostat operates the auxilliary heating system to boost the water temperature. This can have the disadvantage that on some occasions very hot water from the solar cylinder is directed into the bottom of the conventional tank which may be full of cold water. At such times the whole contents of the cylinder will have to be run-off at the taps, before the hot 'solar' water can be used.

To avoid such wastage of water, or heat, a 3—way diverter valve was added to the outlet of the solar cylinder. Inside the valve a gate directs the solar heated water either to the bottom of the conventional hot water cylinder as normal, or it is directed into the outlet from the conventional cylinder and hence direct to the hot water taps. This could have been done manually with two stop-cocks, but in this case I installed the Honeywell V5067 D.H.W. diverter valve, which can be controlled automatically by a T5053E thermostatic actuator and costs about £13 (trade price). This screws on to the head of the valve and is connected to a temperature sensor which is attached to the top of the solar cylinder. The

sensor mechanically activates the diverter valve whenever the water in the solar tank is hot enough to be sent direct to the taps. The valve must be located fairly close to the sensor on the solar cylinder, within 1.5 m (5 ft.). If a greater distance is necessary, more expensive electrical controls will be needed.

When the installation was complete, I remember well my disappointment on turning on the hot tap . . . nothing happened. The problem was soon solved however. Air in the system was causing a blockage and this was removed by applying mains pressure. A small length of hose pipe was used to link the nozzle of a hot tap to that of the nearest cold tap which was connected direct to the mains (i.e. not to the cold storage tank). The hot taps were opened and then holding tight to the hose pipe I turned on the cold tap. After a minute I turned off the pressure and disconnected the hose. With a great deal of gurgling and spluttering water began to flow.

The final mistake showed up after the system had been operating for several weeks. One day the pump controller refused to switch the system off. I first noticed the malfunction after sunset. After discounting the possibility that the system was operating on cosmic rays and moon-glow, I decided that it was an electronic fault. Later however I recalled that the fault had shown up just after some heavy rain.

The connecting cable linking the temperature sensor on the collector to the controller had been too short and had to be extended by an additional length of cable. As I eventually guessed, this electrical join had not been sufficiently well weather-proofed and the rain fall had caused a short-circuit which kept the pump running continuously.

As I write this, in a cold but bright November in 1977 this system is still producing tank-fulls of water heated to about 30—35 deg.C. (86—95 deg.F.) on most days of the week. The performance is being continuously monitored with the assistance of the Science Research Council and with the cooperation of the London Borough of Islington and it is intended to gather sufficient data to verify or improve our methods of theoretically predicting solar system performance. Already it has served the purpose of demonstrating to a great number of visitors the practicality of using solar energy in ordinary homes and surveys we have carried out in the rest of the borough indicate that similar systems could be installed in the vast majority of existing housing.

### Solar Air Heater

Probably the most original solar water heating system I know of is that built by Mr. S. Pallis on his own home in Kent. By now you will have realised that one of the principal reasons for the expense of solar installations lies in the necessity, in most designs, of using materials which are good heat conductors, leakproof and do not corrode. In this system, these requirements are sidestepped by designing a solar collector which heats air instead of water. The heated air is then circulated round a cold water tank to pre-heat its contents before they are delivered to a conventional hot water cylinder. Air is not corrosive, does not cause damage when it leaks and, as it brushes across the whole interior surface when it is circulated through a solar collector, virtually

any dark coloured material can be used as an absorber, there being no need to conduct absorbed heat to specific flow channels.

In this example, black painted insulating board serves as the absorber plate as well as the insulant. In other air heaters, corrugated metal, blackened sheets of glass, and even black muslin have been successfully used. The important criteria for an air heating absorber are that it should distribute the air flow so that there are no dead areas, and that it should encourage turbulent air flow. Turbulency increases the effective contact between absorber and fluid and hence the heat transfer. It is encouraged by providing rough surfaces in the absorber and by placing baffles in the circulation route.

Mr. Pallis used an ingeniously simple method for checking the flow through his collector. He made turbulence-indicating-flags, human hairs glued to the top of a pin which was stuck into the insulating board. These could be observed through the collector cover and changes in the flow were effected by rearranging the pattern of baffles formed by strips of insulating board and sheets of glass.

The collector is arranged below the level of an existing galvanised iron cold storage tank so that the system would operate automatically, with convection bringing heated air up to the tank and the circulation cutting out when there is insufficient sun just as in a gravity circulating water system. The tank has been encased in insulating board, and, between the tank walls and the insulant, corrugated galvanized sheets serve the dual purpose of heat exchange fins and duct formers for the warm air. Using the existing tank in this way makes for a

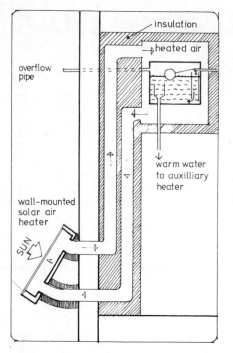

insulation

heated air

overflow pipe

warm water to auxilliary heater

wall-mounted solar air heater

SUN

*15.5 Solar air-heating collector pre-heating a water tank.*

cheaper system. (In most new houses, and many old ones, this would unfortunately not be possible as the cold water storage tank usually feeds cold taps and W.C.'s, as well as the hot water cylinder). The collector is hinged to a south facing wall so that its tilt angle can be changed from season to season in accordance with the sun's changing altitude. This meant flexible ducting had to be used to connect the panel. A 125 mm. (5 ins.) diameter, neoprene coated, woven glass fibre wound around spring steel wire was selected as a material able to withstand the temperature of the solar heated air. A similar vinyl coated tube, 275 mm. (9 ins.) in diameter, was used to weather-proof the insulating lagging which was stuffed around the inner tube. Inside the house, cheaper galvanized steel pipe was used for ducting.

Mr. Pallis, who is an engineering consultant, paid great attention to detail in his design and so avoided many of the errors which a less cautious innovator would have made. Inside the insulated tank casing for example, is a metal drip tray with a drain to discharge the condensation which occurs on the tank's cold surfaces during winter. Inside the tank he installed baffles to prevent the incoming cold water gushing out of the ball valve and mixing with the hottest layers of water at the top of the tank.

At a later date, a two speed fan was introduced to the system to increase the air circulation. The flow rate in air systems can have a significant effect on the absorber's efficiency. Water is about one thousand times denser than air and its specific heat is four times greater. Therefore to effect the same heat transfer between absorber and fluid as would occur in a water heating absorber, with the same temperature rise, an air heater would require a volumetric flow rate some four thousand times greater. To circulate such large quantities of air requires more electrical energy than would be used by a pump circulating the equivalent quantity of water, and it would require inconveniently large ducts. Air heaters therefore, usually work at lower efficiencies than water heaters. The required air/water heat exchange at the storage tank is also a source of inefficiency (unless a more expensive system with finned tubes and circulating water is adopted).

Given these disadvantages, Mr. Pallis concluded that for general applications the more conventional water heating systems offered greater potential. His own system however is probably unbeatable as regards material cost and this of course is often the determining factor behind many decisions relating to heating.

One of the most important lessons arising from Mr. Pallis' work, is the value of reflectors. On realising that the 1.67 sq.m. (18 sq.ft.) collector was not providing as much solar energy as his family could use, he compared the cost of buying an additional commercial collector with the material cost of a reflector which would deflect rays of sunshine which would normally be absorbed by the surroundings, into the collector already installed. For the same boost in annual energy capture, the reflector seemed the cheaper solution.

He built a timber frame in the form of a section of a parabolic cylinder positioned to focus on the collector. This was initially covered with several types of material; 'mill-finish' aluminium, brushed aluminium and an aluminium film vacuum laminated on to PVC. The mill-finish tarnished within three months but the

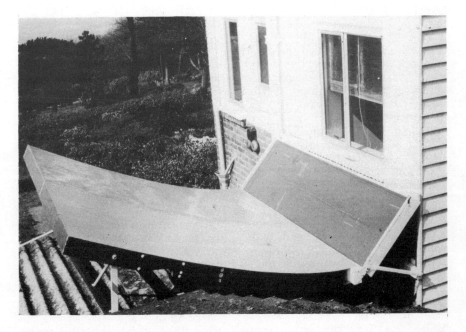

*15.6 Solar air heater with variable pitch reflector.*

film material, which started out as the brightest reflector, showed no deterioration after eighteen months and was later used to cover the entire reflector. This reflector placed below and to the south of the collector actually concentrates the sun's rays due to its shape, and its tilt angle must be adjusted four times a year. Above the collector is a stationary, flat reflector — the wall of the house, painted bright white. Such a reflector is so cheap that it is obviously cost effective. A flat reflective surface, which could be a wall or roof, angled at about 45 degrees from the plane in which the face of the collector sits can make a considerable contribution to water heating. A large reflector can provide as much as 15% of the energy the whole system yields. This is worth remembering, no matter what kind of solar collector you are using.

## Farm Installation

Brian Ford (who illustrated this book) and Nick Moore installed this system on a Somerset farm. The collector is another example of old central heating radiators being put to good use. They were mounted inside insulated boxes made from 50 x 100 mm (2 x 4 ins.) framing with a marine ply base and placed directly on top of the rafters of the farm roof after the tiles had been removed. The radiator and frame dimensions were chosen to accommodate a standard 1410 x 730 mm (4'–7½" x 2'–4¾") 'Dutch Light' (horticultural glass which can be obtained from nurseries for half the price of ordinary window glass).

The connections between the radiators were made with radiator hose and jubilee clips. The radiators provide an absorber area of 7 sq.m. (71 sq.ft.).

Plumbing was carried out with polythene pipe (alkathene). The copper olive and jubilee clip type of connection described in ch. 8 — backyard plumbing techniques — was used initially. These were later replaced with compression fittings however due to numerous leakages.

This is a single-tank indirect system. That is to say hot water from the solar panel is pumped directly to the existing hot water cylinder. As the existing heat exchange coil was connected to an oil-fired Aga with a back-boiler, it was necessary therefore to introduce a second heat exchanger for the solar input. This was done quite easily by removing the electrical immersion heater from the top of the tank and inserting a 'Micraversion' heat exchanger which is manufactured to fit the threaded immersion boss.

Unfortunately part of the new heat exchanger is situated in the top of the cylinder — the hottest region. This makes heat transfer more difficult and has the danger that sometimes the system could operate in reverse. On dull days when the Aga heats the cylinder, the warmth will be conducted into the water in the top of the Micraversion and this could then rise in a convection current up to the cold collectors thereby throwing heat away. In this case, the pump was located on the flow pipe, the outlet from the collectors. When not in use, it acts as a non-return valve thus reducing the likelihood of reverse circulation. It would however be more efficient in terms of heat transfer if the Aga were connected to the Micraversion, and the solar panels to the coil in the bottom of the tank.

The pump inititially installed was one salvaged from a central heating system on the farm. Soon after it was switched

*15.7–15.9 Removing roof covering and installing inset solar collectors on a Somerset farm.*

on however, it seized up. This often happens with old pumps which have been left dormant. Sometimes however people have similar problems with new pumps and this is usually due either to switching the pump on before it is submersed in water, or due to debris in the system blocking the impeller. Pipework and collectors should always be flushed out by mains pressure water before the pump is fitted in order to remove copper shavings and lumps of solder and, where old radiators are being used, scale and rust.

In this example, the Aga back-up system was already there. It is worth noting however that the Aga range, whilst popular in farmhouses for cooking, is not the ideal complement to a solar system. This is because, no matter how much the sun has heated the water, the Aga will always burn the same amount of fuel. This means that on days when both the solar system and the Aga have been operating, the water may become very hot indeed making the efficiency of solar collection low. Energy will be saved of course because smaller quantities of hot water will be used as they will need to be diluted with cold. It would be more efficient however if the back-up heater regulated itself, perhaps with dampers controlled by a thermostat in the boiler. In "Technological Self-Sufficiency" Robin Clarke recommends the use of a solid fuel stove with a glass door, such as the Rayburn Rhapsody 301 or the Parkray 88Q or 99Q. Robin Clarke's own solar heating system is also of interest as it is one of the few examples of a trickle collector. It is described in the next section.

## Trickle Collector System

Trickle, or open trough solar collectors have always appeared attractive on grounds of their low cost and the self evident simplicity of their operation. The first solar collector I ever heard of was a trickle type built by an American pioneer, Harry Thomason. I remember that I had previously associated solar energy with something used in space ships and therefore expected some intricately complex piece of technology when someone began to describe Thomason's solar collector. Our modern world often does train us not to see the wood for the trees and it took me some time to realise that pouring water over a hot corrugated roof was all there was to it.

Although the principle is straight-forward, one obviously runs into snags when applying it. Robin Clarke, with advice from the architects Robert and Brenda Vale, has now built two such solar roofs and has therefore worked through many of the practical problems.

On his own house, he removed the old slates from a 16 sq. m. (170 sq. ft.) area of south facing roof, and replaced them with a layer of chipboard covered with roofing felt then decked over with corrugated aluminium annodised dark brown. 100 mm. (4 ins.) of insulation was added between the rafters. The aluminium is fixed with aluminium nails with plastic washers and pop seals, nailed through the ridges and not the channels.

A 22 mm. (¾ ins.) copper pipe was located along the roof ridge and 4.8 mm. (3/16 ins.) holes drilled along its length so that when the pipe is filled, water

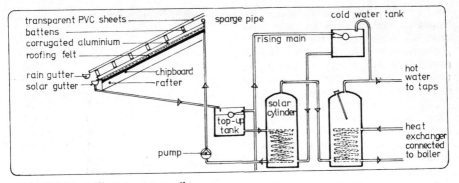

**15.10 Trickle collector system diagram.**

pours out into the channels formed by the corrugations. The ends of this sparge pipe are capped with compression fittings which allow access for cleaning the interior should blockages ever occur. The pipe is connected to a pumped water supply, not in the centre but from two points, a quarter way along from each end, in order to spread the flow more evenly across the roof.

Along the lower edge of the aluminium, a plastic gutter pipe was installed to collect the water running down the channels. "If you possibly can" said Robin Clarke, "make sure this gutter is outside the house."

In the first solar roof he built there were problems with the gutter overflowing as the water was not running out of it fast enough. He advises using a pipe larger than 28 mm. (1 inch) in diameter for leading the heated water from the gutter to the tank inside.

The water is not delivered directly to the hot water cylinder. First it is discharged through a fine mesh kitchen sieve to filter out any debris it has collected on its journey over the roof, then into a 45 litre (10 gallon) open top tank. This tank is connected to the

mains with a ball cock in order to top up the system when evaporation losses make that necessary.

From the top-up, the water is pumped into the heat exchange coil in a 150 litre (30 gallon) copper cylinder. Having given up its collected heat, it is then pumped back to the roof top for reheating at a rate of no less than 8.5 litres/hr/m$^2$ of collector (0.19 gall/hr)/ft. The pump, a 180 watt circulator, in this case, must be located well below the water level in the top-up tank to ensure that it is always submerged.

The collector was covered with ICI Novolux. This is a corrugated transparent PVC which can be screwed directly onto timber battens laid across the aluminium. One problem encountered with this material was due to its low softening temperature. At the ridge, the plastic was originally overlapped with roofing felt. The currents of warm air rising to the ridge, however melted the bitumin in the felt which then formed an air-tight seal with the plastic. This allowed the internal temperature at the ridge to climb even higher and eventually the Novalux softened and sank below the sparge pipe, redirecting the pumped water over the top of the plastic sheeting and into the rain water gutter. This problem was overcome by replacing the felt with a lead flashing at the ridge. This dissipates the convected heat more rapidly, lowering the efficiency of solar collection, but allowing an economical and convenient material to be used in place of glass.

The whole installation was carried out by two people in the space of one week. The cost breakdown was as follows (1975 prices):

| | |
|---|---|
| aluminium | £40 |
| Novalux | £35 |
| chipboard | £12 |
| roof felt | £ 3 |
| gutter | £ 4 |
| top-up tank | £ 8 |
| cylinder | £27 |
| pipe & fittings | £ 8 |
| pump | £17 |
| auto pump switch | £20 |
| hard wood battens | £ 2 |
| nails & screws | £ 3 |

The total cost of the solar installation, excluding, that is, the back up system, was £179. It also saved £120, the estimated price a builder would have charged for renewing the slate roof. During the first three months of summer operation, it saved an estimated £14.40 worth of electricity. Robin Clarke describes his experience of putting solar energy to use, along with all the other adventures involved in rebuilding a tiny cottage into a home for a community of sixteen people in his book, "Technological Self-Sufficiency". Those who have considered taking a more active role in shaping their personal environment, but doubted their ability, will find reading this a real tonic.

## School Swimming Pool Heater

The 'Resources' group in Welwyn Garden City demonstrated very effectively that environmentally concerned organizations can do much more than just talk. They mounted 38 sq.m. (412 sq. ft.) of polypropylene solar collectors on the roof of a local school to heat the outdoor swimming pool. Coordinated by Tony Wigens of Country College, they persuaded several private companies to donate materials and then did the necessary plumbing wiring and constructional work with their own volunteer force.

The collectors are mounted on a frame work of prefabricated tubular steel and linked with flexible connectors, jubilee clips and PVC pipe. The collectors are connected in two banks of eight. In the system diagram it can be seen that the heated water leaving the first bank is led in a 38 mm (1½ ins.) pipe to the end of the second bank where it is tee-ed into its outlet. The united flow is then carried in a 50 mm (2 ins.) bore pipe back to the pool. It would be easy to imagine alternative arrangement which would reduce the length of piping needed, avoiding having the flow from the first bank doubling back on itself. The configuration shown however was recommended by the collector manufacturers in order to ensure a more even flow rate through the two banks.

The pool was previously heated by pumping its water through a large steel cylinder known as a calorifier. Inside this it was heated by a heat exchanger linked to a boiler. When adding the solar heating circuit, the pipe bringing pool water into the calorifier was broken into, and

154

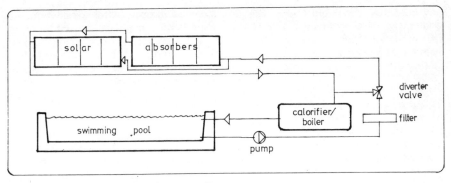

*15.11 School swimming pool system diagram.*

a motorised, three-way diverter valve installed. This is like a tee-piece which has a moveable gate inside which will direct the flow through one or other of the two branches offered by the tee. In this case one branch would lead the pool water directly through the calorifier as in the past. The alternative branch leads to the solar collectors. An electric motor operates the diverter gate and this is controlled by a roof mounted sensor which measures the sol-air temperature. The water therefore is only pumped through the collectors when weather conditions suggest that there will be a useful heat gain.

Caution needs to be exercised in carrying out the electrical wiring in the wet environment of a swimming pool and all precautions must be taken to protect

*15.12 School swimming pool heated by solar collectors mounted on roof over classrooms.*

any pool user who might touch exposed equipment with wet hands. If cable is laid externally and is liable to physical damage, stranded wire armoured P.V.C. cable could be used as in this case.

This installation is an excellent example of what direct action by local groups can achieve. I would hope that it serves as an inspiration to others, such as parent-teacher associations, who are looking for a worthwhile group activity. It might also make an interesting 'live' project for older pupils in science and technical classes. After installation the solar system could serve, not only to heat the pool, but also as an educational tool serving as an introduction to ideas of environmental protection or as an experimental apparatus for science classes who could monitor its performance.

Temperature readings were taken by the Welwyn school and the results recorded in the first few days after its installation are reproduced below:

### Solar Heating Temperatures

B = School boiler still functioning

| Date | Time | Air temperature | Water temperature | Weather conditions | Solar heater on/off |
|------|------|-----------------|-------------------|--------------------|--------------------|
| 27.6.77 | 3.00 | 19°C | 25°C | Overcast (B) | Off |
| 28.6.77 | 9.00 | 15°C | 22°C | Rain (B) | Off |
| | 12.00 | 16°C | 24°C | Overcast (B) | Off |
| | 3.00 | 22°C | 25°C | Bright Periods (B) | On |
| 29.6.77 | 9.30 | 18°C | 25°C | Sunshine (B) | On |
| | 12.30 | 19°C | 27°C | Sunshine (B) | On |
| | 3.30 | 17°C | 27°C | Bright Spells (B) | On |
| 30.6.77 | 9.30 | 18°C | 25°C | Sunshine (B) | On |
| | 1.30 | 19°C | 27°C | Bright Spells (B) | On |
| | 3.00 | 19°C | 27°C | Cloudy (B) | On |
| 1.7.77 | 9.00 | 17°C | 23°C | Cloudy | Off |
| | 12.00 | 17°C | 23°C | Cloudy/Bright | On |
| | 1.30 | 17°C | 24°C | Cloudy | On |
| 2.7.77 | 10.30 | 19°C | 22°C | Sunshine | On |
| | 11.15 | 19°C | 22.5°C | Sunshine | On |
| | 12.15 | | 23.5°C | Sunshine | On |
| | 2.45 | | 25°C | Sunshine | On |
| | 4.00 | 19°C | 25.5°C | Sunshine | On |
| 3.7.77 | 10.00 | 19°C | 24.5°C | Sunshine | On |
| | 3.00 | 22°C | 26°C | Sunshine | On |
| 4.7.77 | 9.00 | 22°C | 24°C | Sunshine | On |

*15.13 Resources' volunteer tightens the final hose clip, securing the inlet pipe to the solar collector array.*

Between 1974 and 1976 there was a very special event held each summer in Bath — Comtek, a festival of community technology. It was organised by the local Arts Workshop and participants and visitors came from all over the country and abroad.

This solar water heater, built on site to heat the water for the festival kitchen is an example of the activities which took place. It served as a working example of how simply and cheaply we could benefit materially from the sun's energy.

*15.14 Solar water heater for festival kitchen being filmed by Granada TV.*

Second-hand radiators are again the basis of the system. They were mounted on a row of doors, also salvaged from demolition sites, with a single layer of insulation quilt between them. The doors were tied to a scaffolding frame and polythene sheeting was stretched over the whole assembly with spacer strips holding it above the radiator panels. Old rubber hose from automobile engines was used as connector pipe between the panels and polythene piping linked the collector to a galvanised tank mounted overhead to make thermosyphonage possible. Given the proximity of the Comtek depot, with its large store of recycled building materials, it was possible to construct this system without spending more than £30.00. It took four days to assemble. One of the pleasing aspects of the construction of this system was that many people contributed to it, including individuals who had come with the idea of being passive onlookers at an exhibition.

15.15 Festival solar water heating system with plastic glazing sheet drawn back.

In situations like this where the primary emphasis is on low first costs, and where long life is not an important requirement, availability of materials determines what is used to a much greater extent than ideas of optimum system design. It is worth saying though that leaks do become very frustrating. They usually occur at joints and it is therefore worth spending a little extra money in order to ensure that there is not too great a variety of pipe sizes and types being used. This variety which tends to occur when using materials closest at hand makes the sealing of joints very difficult.

Airlocks caused considerable delays in getting the system working. The use of rubber and plastic connecting pipe, however, made the job of shaking and rocking the air out much easier.

159

Table 16.1 Characteristics of some commercially available solar collectors.

| Manufacturer | Plate | Casing | Insulation | Glazing | Plate Surface | Capacity $Lit/m^2$ | Plate Area $m^2$ |
|---|---|---|---|---|---|---|---|
| Alcoa | alum. roll-bond | none | none | none | stove enamelled | 3.0 0.9 | 2 1 |
| Air Distribution | stainless steel tube alum. plate | fibre-glass | 25mm poly-urethane | glass | black paint | 1.2 | 0.37 |
| Consumer Power Co. Ltd. | pressed steel rad. | formica alum. steel mounts | 50mm glass-wool | 32oz. glass | black paint | 12.3 | 1.08 |
| Distrimix Ltd. | steel tube & steel plate | galv. steel | 50mm rock wool | none | electro plated "selective" black | 2.97 | 1.43 |
| Don Engineering | pressed steel radiator panel 18gauge | stainless steel | 25mm urethane foam | none | stove enamelled "selective" paint | 6.9 | 1.64 0.78 |
| Drake and Fletcher | alum. tube & alum. plate | alum. | 19mm exp. polystyrene | none | etched and black paint | | 0.88 |
| A.T. Marston, Calorsol Ltd. | | | | | | | |
| Mk.III | pressed steel radiator panel | G.R.P. | 25mm foam | transparent G.R.P. | black paint | 9.47 | 2.4 |
| Mk.I | Drip-feed corrugated alum. | G.R.P. | 25mm foam | transparent G.R.P. | black paint | — | 2.4 |
| Production Methods Ltd. | | | | | | | |
| Mk.I. | copper tube plate | alum. | 50mm. min. wool | none | black paint | 1.47 | 1.44 |
| Mk.II | alum. tube alum. plate | alum. | do. | do. | do. | do. | do. |

| Manufacturer | Plate | Casing | Insulation | Glazing | Plate Surface | Capacity $Lit/m^2$ | Plate Area $m^2$ |
|---|---|---|---|---|---|---|---|
| Robinsons of Winchester | | | | | | | |
| Mk.I. | 4mm thick P.V.C. wafer | alum. | none | none | black paint | 4.18 | 2.4 |
| Mk.II | Poly-carb. wafer | alum. | | | black paint | | 3.6 |
| Solar Centre | | | | | | | |
| Sunstor MK.I | copper tube copper plate | alum. | 50mm glass-fibre | none | black paint | 0.71 | 0.9 |
| Sunstor Mk.II | corru-gated alum. | alum. & plywood | urethane foam | none | baked on paint | – | 1.2 |
| Solar Heat Ltd | copper tubes galv. plate | corru-gated plastic | 25mm glass wool | 24oz glass | | | 1.02 |
| Solar Water Heaters Ltd | poly-prop. wafer | | 63mm exp. polystyrene | double poly-carb. | black plastic | | 3.3 |
| Stelrad Radiators | pressed steel rad's. | none | none | none | none | 5.0 | 1.36 |
| Sun Heat Systems Ltd | abs. plastic sand-wich | alum. + oil imp. hard-board | 30mm fibre glass | 4mm glass | black plastic | 3.36 | 0.75 |

NB See appendix 6 for a more complete list of solar collector manufacturers.

# 16

# Buying Solar Collectors

Whether you are going to build your own solar collectors or buy them ready-made, it is worthwhile taking the opportunity to examine a selection of those which are on the market. In addition to the occasional trade exhibitions, there is a permanent display featuring a large number of different collectors at the National Centre for Alternative Technology at Machynlleth in Wales.

It is a sign of the increasing interest in solar technology that there are now more than sixty companies in the UK manufacturing solar collectors. As in any other expensive purchase, it is important that you shop around and do not jump at the first offer from the growing army of door-to-door solar salesmen.

There are as yet no official standards to which solar collectors must conform. Nor is there an agreed testing procedure which would allow us to compare the performance of one model with another. The performance data which some companies do present may or may not help to indicate what energy returns the

collector will provide, but it will not be of use in comparing with the data of another company who may have performed similar tests but under different conditions. Until official tests are prescribed and a standard method of presenting the results is agreed, performance figures might as well be ignored when comparing the products of different companies. At present we must rely upon broad generalisations based upon the basic physical principles outlined in ch. 3. It is important to remember therefore that the advice which follows is applicable to the class of collector being described but possibly not to an individual collector within that class. What I am attempting to do, is indicate the characteristics you ought to examine in any particular collectors you wish to compare.

## Prices

This is the most obvious of differences between different manufacturers. Make sure however that you compare like with like. The cost/panel is for example no basis for comparison when panels vary from 0.36 sq.m (4 sq.ft.) to 3.6sq.m. (39 sq.ft.). Always ask for the unit cost, that is the cost/sq.m. (or cost/sq.ft.) This can be found by dividing the cost/panel by the absorber plate area.

$$\text{cost/sq.m.} = \frac{\text{cost/panel}}{\text{absorber area (sq.m.)}}$$

Note that you should use the absorber area and not the total panel area which includes framing and glazing strips which do not contribute to the actual collection of energy.

This may seem straight forward and obvious but I remember the first time I examined the market I became confused by the difference between the cost/sq.m. quoted to me by a company and the cost/sq.m. which I calculated from the panel price and the absorber area.

This particular company was marketing an adapted central heating radiator as a solar absorber. In calculating the heat emission area for central heating purposes, the entire surface area of the radiator is taken account of. This then includes all the corrugations and mouldings on its surface so that a panel measuring 1 m. x 1 m. does not have an upper surface of 1 sq. m. as you might expect, but something like 1.2 sq. m. This company had done a similar calculation with their absorber.

With solar collectors however, the quantity of solar energy intercepted will be the same no matter how smooth or uneven the surface of the absorber is. To include the area of surface irregu-

larities is therefore highly misleading particularly as it makes the cost/sq.m. appear lower than it really is. It is worth taking a note of the absorber plate dimensions if you want to make a thorough appraisal of the collectors available, then you can check the given figures for absorption area and unit cost, and see if you are dealing with a wily salesman.

A second point to note when scrutinising cost is what exactly is being offered for the price being asked. £60 sq.m. is a fairly standard price for a domestic solar collector. The installed cost however, which includes ancilliary equipment and fitting, is more likely to be £100–£150 sq. m. Even when dealing with the price of the collectors alone, there is considerable difference in what is being offered. Swimming pool collectors are often uncased and unglazed absorber plates whereas domestic water heaters need insulation, glazing and casing. The quoted prices however do not always include glazing and this would mean a necessary additional expenditure of about £5–£10/sq.m.

### Design

Apart from a few unusual designs which are described later, all the commercial collectors can be classified in the three categories suggested in the chapter on solar collectors. Namely:

(i) trickle or open trough absorbers
(ii) sandwich absorbers
(iii) tube-and-sheet absorbers

These are illustrated in ch. 4 and it is well worth reading that whole chapter before going shopping.

Tube-and-sheet designs are probably the most effective for operation in cloudy climates. They have the advantage of having a low thermal capacity and hence can react more rapidly to short outbreaks of strong sunshine. The sheet must however be made from materials which are good conductors of heat, usually copper or aluminium, and these tend to be expensive.

Sandwich designs have the potential advantage of being cheaper both because they can be made from a wider range of materials as heat conductivity is not such a critical characteristic, and because the form is one which lends itself more easily to mass-production techniques. If their water capacity, and hence their thermal capacity, could be kept low, they would give a performance comparable to that of the tube-and-sheet models. At present however only the aluminium Roll-Bond variation of the sandwich design has attained water capacities comparable with tube-and-sheet models (i.e. less than 3 litres/sq.m. (0.065 galls/ sq.ft.)). The Roll-Bond absorber is in fact a patented design which falls between the two classifications, being like a sandwich absorber in construction and like a tube-and-sheet absorber in performance.

Although water capacity has been recognised for some time now as a critical factor affecting performance, I have yet to see published data of any comparative experiments carried out in this country which would enable one to quantify the overall effect.

Trickle collectors are much less efficient than either of the other types. This coupled with the simplicity of their construction makes me doubt whether it is worthwhile buying a manufactured trickle collector for domestic applications.

## Materials

The ideal choice of materials will be different for different collector designs. What one is looking for is the materials which will cost least, perform their function efficiently and last considerably longer than the pay-back period of the collectors. Taking account of the functions of the different types of absorber plate constructions described in ch. 4., we might say that the most suitable materials for solar absorbers are as follows:

trickle absorbers —
    aluminium, annodised or with a stoved-on coating.
sandwich absorbers —
    steel (to be used only in indirect circuits).
    polypropylene or polycarbonate (used in unglazed collectors for low temperature applications such as swimming pools).
tube-and-sheet —
    copper or stainless steel tubes with copper or aluminium sheet.

Note that aluminium is not recommended as a suitable material for any part of the collector in which it will be in direct contact with the circulating stream of water. This is because of the possibility of early problems with leakages due to corrosion. In spite of this, there are many companies selling collectors with aluminium waterways.

### Casing

If you are buying your collectors, you should choose one with a durable casing. Aluminium and rigid fibre glass are good light-weight materials for this purpose. Timber casings, although recommended for DIY collectors because of their ease of production, will only prove durable if they are given regular protective coatings.

## Glazing

Glass is the best choice for a long-life, high transmission, collector cover. Plastics do however offer the convenience of factory-finished collectors which can be transported easily without damage. Double-glazing does not appear to be cost-effective in most cases.

## Insulation

Some insulation is required on collectors intended to heat domestic hot water. Many materials can be successful but avoid collectors using polystyrene. This can, and often has, melted due to the temperatures achieved inside solar collectors and manufacturers who do not realise this have not done their homework.

## Unusual designs

In addition to the collector types already described, there are four others on the market, concentrating collectors, heat-pipe collectors, coil collectors and combined absorber/storage collectors.

Concentrating collectors for heating water usually take the form of a polished metal mirror in the form of a cylindrical trough with a blackened absorber pipe running along its focus. Depending upon the geometry of the curved mirror, the collector may be stationary, or tracking. The tracking concentrators are more efficient and achieve higher temperatures. They incur however the additional cost of a tracking motor in order to follow the sun. The greatest disadvantage of concentrating collectors is their dependence upon direct sunshine. They cannot concentrate the diffuse radiation which comes through cloud cover. More than half of the solar energy received in a country like Britain is therefore lost to concentrating collectors. It has been suggested that an annual total of 1500 hours of bright sun is necessary to make concentration of the sun's rays worthwhile. In the U.K. we receive only about 1200 hours of bright sun each year.

Heat-pipes are sealed and partially evacuated tubes with a capilliary mesh material lining their interior walls. Their name derives from their ability to transfer heat along their length very efficiently. This is done by means of a working fluid which is injected into the tube prior to evacuation. Heat transfer is effected by an evaporation—condensation—evaporation cycle. When heat is applied at one end of the pipe, the fluid evaporates absorbing the applied heat as latent heat of vaporisation. The vapour is then transported to any cooler area of the tube due to pressure differentials and is then condensed on the cooler area of tube wall. The condensed liquid is then absorbed by the mesh lining which returns it to the heated area by capilliary action.

Solar collectors have been manufactured which utilise heat pipes to replace the conventional waterways on a tube-and-sheet absorber. These have the advantage that the need for frost protection is obviated as the exposed collector need contain no water. Heat can be transferred to water in an insulated water jacket around one end of the heat pipe. As yet, heat pipe collectors have not been shown to have a higher efficiency than can be obtained with more conventional designs. Their frost-proof advantage however makes them an interesting alternative particularly if their cost can be brought down to the level of ordinary collectors.

Coil collectors are simply two dimensional spirals — a length of pipe coiled in concentric circles of ever decreasing diameter. Water is pumped in at the perimeter and exits from the centre. Its chief advantage is a reduction in heat losses due to the hottest part of the absorber, the centre, being partially insulated by the surrounding perimeter coils. If the concentric circles are in edge contact, construction is very simple but material costs are high. Sometimes the coils are spaced out and attached to a backing sheet of copper or aluminium. This can then be seen as a variation on the tube-and-sheet type of absorber.

Combined absorber/storage collectors are now being imported from Japan for sale in this country. They consist of large water vessels inside insulated casings. By combining the functions of collecting and storing energy in one element, initial costs can be cut; there is no need to buy an additional storage tank. In very sunny regions this approach is a valid way of providing solar water heating at a lower cost. Their prime disadvantage lies in the high heat losses which occur due to having one side of the storage vessel poorly insulated with only a sheet of glass for cover. A second disadvantage is that as heated water remains in the collector, its operating temperature will rapidly rise and hence lower the efficiency of energy collection. Ch. 3 explains the advantages of separ-

ating absorption and storage. Finally the units are much heavier than conventional collectors and additional structural support may be necessary when mounting them on roofs.

## Solar cowboys

In a rapidly expanding new industry it is inevitable that there will be a few shady operators in the business for quick profits, and with little understanding of their subject. Without first familiarising yourself with some of the basic principles being employed and the likely yields a solar system can provide in terms of energy and finance, it is not always easy to recognise the people you want to avoid. Certainly the usual signs of respectability, a big company and staff with strings of qualifications, are not always a guarantee of technical competence.

Perhaps the easiest way of recognizing rogue companies is through a close examination of their advertising literature. Do not be impressed by glossy pamphlets but pick out any definite details such as specification of materials used in construction, predictions of system performance and pay-back periods. Details such as these should be compared with recommendations and figures given in this book (or other impartial sources such as UK-ISES, the International Solar Society).

Some companies quote the annual total of solar energy received in south England as being about 1000 kWh/sq.m. and imply that each square metre of their collectors will supply this amount of energy, omitting to mention that unavoidable inefficiencies in the collectors will more than half this quantity. Other companies get the figures completely wrong — suggesting that their collectors have an output in excess of 1.4 kWh./sq.m. Even if they were orbited in space at the outer edge of our atmosphere they would not intercept energy at that rate, let alone convert it to useful hot water.

On the earth's surface the rate of receiving solar energy rarely exceeds 1.0 kWh/sq.m. The yearly mean value of irradiance in the UK is about one tenth of this, that is 0.1 kWh./sq.m. of horizontal surface. The annual output of a solar water heating system in the UK is likely to be about 300 to 400 kWh. for each square metre of collector surface. Any claims for yields much higher than this must be viewed with suspicion.

Having given all these warnings, I would like to add that there are indeed a great many technically competent and honest companies operating in the solar energy business and there are sometimes even signs of genuine idealism from those who appreciate the universal benefits of turning to a more sustainable energy source, and see industrial activity as being the only effective means of achieving the necessary change.

## Commercial Products

In 1975 I made a detailed study of all the collectors which I had seen advertised. The results of my enquiries are indicated in the table below. There is no attempt made to pass judgement on any of the products mentioned. Readers can come to their own decisions by comparing the specifications shown with the recommendations which have already been made. It is important to note that there have been many changes in the design specifications and materials used in the two years which have elapsed since the table was compiled.

The table should therefore not be used as a shopping guide to solar collectors, but rather as an indicator of the variety of collectors on the market and an example of the details which should be examined when assessing different products.

# Appendices

In this appendix, John R.G. Corbett describes how to build an electronic temperature differential control unit which he has designed for pumped solar systems. The plans shown form part of a Patent Provisional Specification but he does not intend this to prevent DIY enthusiasts who wish to make one for themselves from doing so.

## Electronic control

The electronic circuit requires a nominal 12 volts d.c. supply. How do we obtain 12 volts d.c. for 24 hours a day, every day? Dry batteries are too expensive. A mains power unit would use about 4 watts; that's 35kWh a year and it's the power we are trying to save!

A rechargeable battery is one answer because the electronic circuit requires only about 10 milliamps, which is 0.24 ampere hours a day. So use your old car battery: it won't even know it's being used because it used to have a capacity of something like 40 ampere hours.

You can connect a battery charger to the battery and its mains lead into the pump so when the pump is on, the battery charger will keep the battery fully charged. Make sure that the battery cannot discharge through the charger when the mains is off.

If you do not want to commit your battery charger to this job all the time, recharge the old car battery as necessary; this may be every few months.

## Black box

The switching on and off of the mains electricity supply to the electric pump is via an electronic black box. But you have no complex electronic circuit to build, because all except one transistor are enclosed in the integrated circuits used.

The black box is necessary continually to sense the panel/tank temperature difference and to start and stop the pump as necessary. The circuit can be as simple or sophisticated as you care to make it. The one described is the minimum required for effective motor control (fig. A1)

A small glass-bead thermistor is attached to the top of the flat plate collector and another to the solar storage tank. This second thermistor on the storage tank should be about halfway up the tank, mid-way between the inlet and outlet tubes coming from the solar panel.

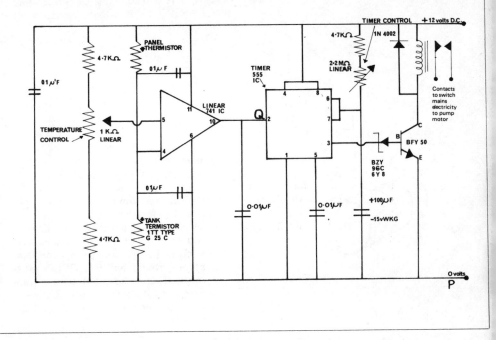

## Varied resistance

The electrical resistance of a thermistor varies with temperature. The higher the temperature, the lower the thermistor resistance. When the panel is hotter than the tank the panel thermistor will have a lower resistance than the tank thermistor so an unbalance will occur at the amplifier input.

This unbalance voltage will be amplified, start the timer which causes the relay to operate, and the relay contacts will connect the mains electricity to the electric pump which now starts pumping. Once it has started, the pump will remain on for a minimum duration of the timer, in my case 100 seconds.

If, at the end of this present duration, the panel temperature is reduced to, or below, that of the storage tank then the pump will stop. If the panel is still 6°F (3.5°C) or more, above the tank temperature, the pump will remain working until the temperatures are equal and then for a further 100 seconds — the duration of the timer.

The theoretical diagram of the electronic circuit is given in fig. A1 and the pin connections on the integrated circuits and transistor are shown in fig. A2. The circuit can be most easily built on printed circuit board perforated with holes spaced 0.1 inch, such as Veroboard. The action of soldering the electronic components to the copper cladding, on the printed circuit board, for electrical

171

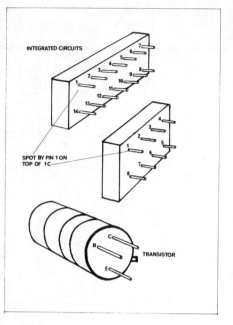

INTEGRATED CIRCUITS

SPOT BY PIN 1 ON
TOP OF 1C

TRANSISTOR

connection also secures them mechanically. The two controls, for temperature and for the timer, should be accessible with a screwdriver.

## Adjustments

When the electronic circuit has been built and before the two thermistors are installed, the temperature control should be calibrated. To do this, first keep both thermistors at the same temperature (around 65°F (18°C) is satisfactory). Then adjust the temperature control until the pump is off. Gradually rotate the temperature control until the pump starts.

Leave the control in this position which can be marked 0°. Now rotate the control in the same direction for another 90 degrees (through a right angle). It is now set to operate the circuit when the panel/tank temperature difference is about 6°F (3.5°C).

The timer control is set when the plumbing is finished and the complete system built, but I will describe how to do the adjustment now.

The action of the relay contacts, in switching off the pump's electric motor, can sometimes cause the electronic circuit to start another cycle of operation. If this does occur at first it can be prevented by locating the relay near the pump motor and the electronic circuit some few feet away.

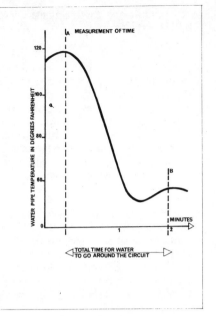

## Take the temperature

Leave the electric pump switched off on a sunny day and wait until there has been at least two hours of almost continuous sunshine. The water in the panel will then have risen to 100°F (38°C) or higher. Attach a thermometer so that it makes good contact with the copper outlet tube of the solar panel.

Wait for the reading on the thermometer to settle down, then switch on the mains electricity and the pump will start. Write down the temperature every ten seconds for at least five minutes and then draw a graph with the results (fig. A3).

The temperature may rise a few degrees to start with. Mark this point A on the graph. Then it will fall as all the hot water leaves the solar panel.

Because the water is being pumped around the circuit all the time, the original hot water will pass through the solar panel for a second time, giving the second rise in temperature some time later. This is point B. The time between A and B is the time it takes for water to be pumped around your complete circuit.

Now measure the length of tube from the outlet at the top of the solar panel to the inlet to the heating coil on the solar water cylinder and call this L1 ft. Measure the length of tube from the heating coil outlet at the bottom of the cylinder to the inlet to the solar panel. Call this L2 ft.

From these measurements you can find the time duration to set your electronic

173

timer. This time will apply only to your system and has to be measured because it varies with size of pipe, type of pump and its setting and so on.

Time in seconds=

$$\text{Total time for water to go round circuit} \times \frac{L1 \text{ ft}}{L1 \text{ ft} + L2 \text{ ft}}$$

**Adjusting timer**

The timer control on the electronic circuit has to be adjusted so that it operates for this time duration. To make this adjustment, wait for a dull day with little sunshine. Then turn the timer control fully clockwise and connect a piece of wire to point P on the electronic circuit. Connect the 12 volt d.c. supply to the circuit and the mains electricity supply for the pump.

Just touch the other end of the piece of wire on to point Q on the circuit (fig. A1). Then remove it and the electric motor driving the pump will start. Note the time for which the pump runs. If this is not the correct duration, rotate the timer control some 90 degrees anti-clockwise, start the motor again and note the time for which it runs.

Now either increase or decrease the time duration with the timer control and run the motor until you get within about ten second of the required time duration. The control can now be left as it is set indefinitely; it will require adjusting only if you change anything in the solar panel water circuit.

The solar radiation which falls upon the outskirts of our atmosphere has a power of 1400 w/sq.m. On passing through the atmosphere however, a considerable quantity of the energy is reflected and absorbed so that on the earth's surface, the sun's power is rarely much more than 1000 w/sq.m. If we take a look at the average amount of power available over twenty-four hours throughout the year, this is of course much lower. In New Mexico, USA and central Australia, it exceeds 250 w/sq.m. whilst in the UK it is a little over 100 w/sq.m. The variation in the availability of solar energy in different locations is shown on the map of the world (fig. A4).

The differences are due not only to latitude, (countries close to the equator generally receive more energy than those further away) but also the local conditions such as altitude, cloud cover and atmospheric pollutants. There is for example a clearly measured reduction in the solar radiation readings taken at the London Weather Centre in the middle of the city, and those from Kew, on the outskirts.

As most people will have noticed, there is also considerable variation in the amount of sun we receive between summer and winter. In the UK on an average summer day we will enjoy something in the order of ten times the amount of solar energy than on an average winter day. These differences can be seen in the maps of the UK (figs. A5–A8) which show the average daily total of solar energy received on a square metre of horizontal surface for four different months.

Fig. A9 shows the annual total of solar energy received on horizontal surfaces of 1 sq.m. in different parts of the UK. This annual total is a useful quantity to know because, as indicated in ch. 13 it can help us calculate, in a simple way, the savings a solar water heating system can make on our fuel bills.

175

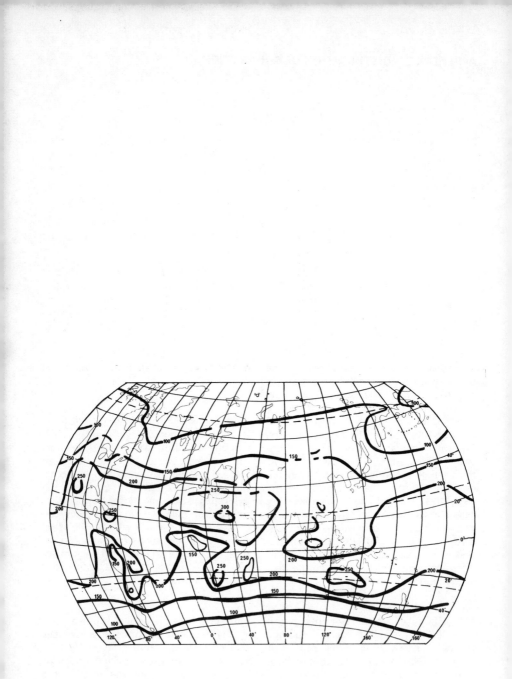

*A.4. Annual mean global irradiance on a horizontal plane at the surface of the earth ($W/m^2$ averaged over 24 hours).*

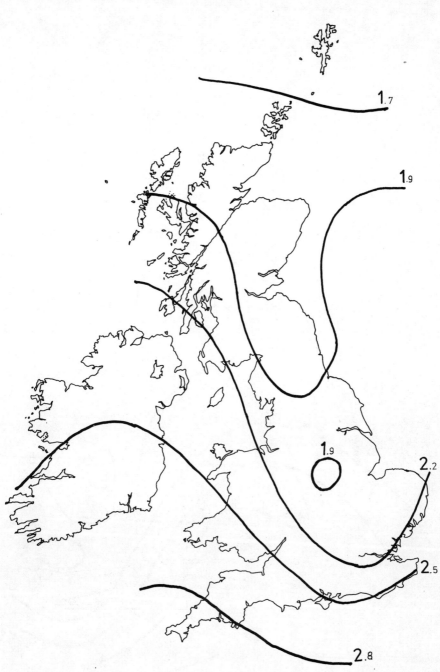

A5. Average daily total of solar radiation received on a horizontal surface in March ($kWh/m^2$).

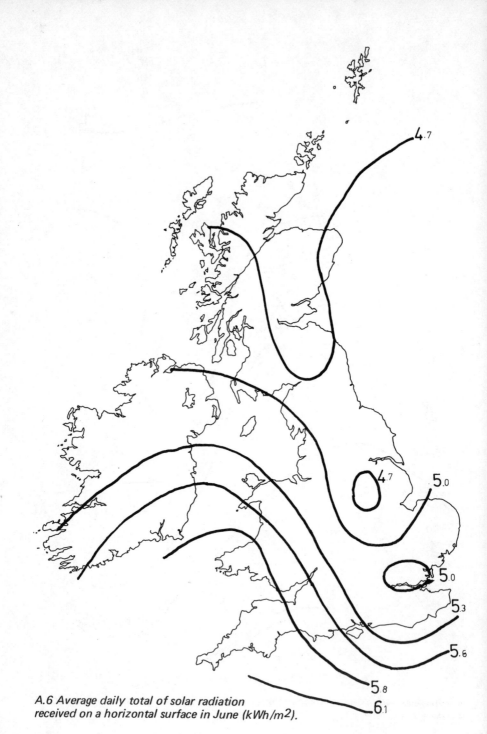

A.6 Average daily total of solar radiation
received on a horizontal surface in June (kWh/m2).

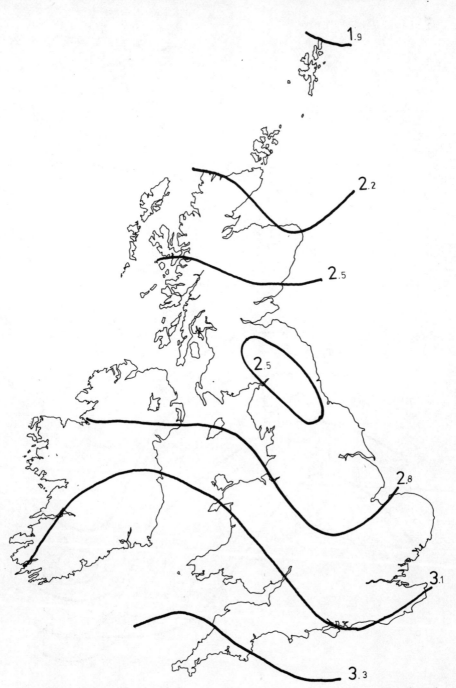

A.7 Average daily total of solar radiation received on a horizontal surface in September (kWh/m²).

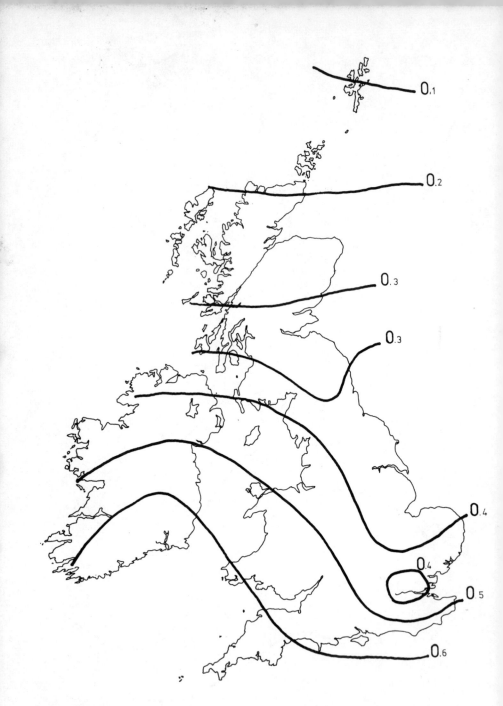

A.8 Average daily total of solar radiation received on a horizontal surface in December (dWh/m2).

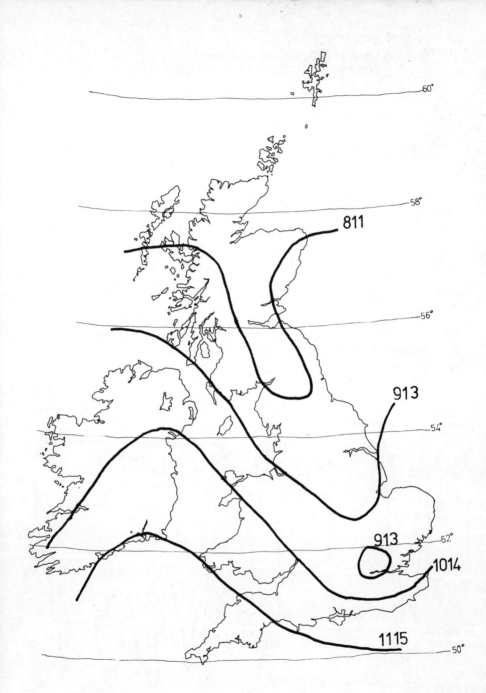

*A.9 Annual total of solar radiation received on a horizontal surface (kWh/m²).*

**Wavelength distribution of solar radiation and glass transmission**

The jagged line in the middle of fig. A10 indicates the amount of energy available in different wavelengths of solar radiation. Values can be read on the vertical scale on the right hand side. Most of the energy can be seen to be concentrated in the visible wavelengths. The lower jagged line shows the amounts of energy which are transmitted through a single sheet of float glass. The upper line shows the transmittance value for float glass. Note that this also varies according to the wavelength. It is at a maximum in the visible wave band and drops to near zero for wavelengths greater than 3000 nanometers. This is very important for solar collectors because all the solar energy arrives in wavelengths shorter than this but as shown in fig. A11, re-radiated heat from the absorber has wavelengths greater than 3000 nanometers.

(figure reproduced by kind permission of Pilkington Bros. Ltd.)

*A.10 Spectral distribution of solar energy transmitted by glass.*

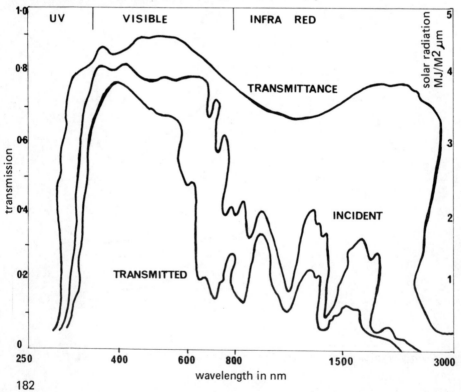

**Glass transmissivity and the greenhouse effect**

This diagram explains why the greenhouse effect occurs. The solar absorber emits heat like a black body. This emission starts at wavelengths greater than 3000 nanometers. This is the same point where the transmissivity of glass drops to zero hence the glass is opaque to the long wave reradiation from the absorber. For this reason, heat losses from a solar absorber can be greatly reduced by covering it with a sheet of glass, and for the same reason the gardener's greenhouse can reach much higher air temperatures than its surroundings.

*A.11 Spectral transmission of glass and emission from a black body at 35°C.* (figure reproduced by kind permission of Pilkington Bros. Ltd.)

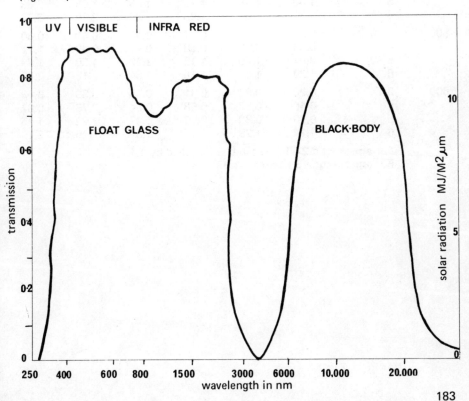

## Glass — Choice of Thickness

These tables prepared by Pilkington Bros., the glass manufacturers, indicate the thickness of glass which is necessary for structural purposes for different areas and support conditions.

### Maximum permitted area for glass squares
### (Aspect Ratio 1 : 1) supported on 4 edges
### (in m²)

| wind load N/m² | glass thickness mm | angle of inclination from the horizontal | | | | | |
|---|---|---|---|---|---|---|---|
| | | 20° | 30° | 40° | 50° | 60° | 70° |
| 1000 | 3 | 0.34 | 0.38 | 0.45 | 0.53 | 0.64 | 0.84 |
| | 4 | 0.64 | 0.71 | 0.82 | 0.97 | 1.18 | 1.58 |
| | 5 | 1.06 | 1.18 | 1.36 | 1.59 | 1.95 | 2.63 |
| | 6 | 1.51 | 1.67 | 1.93 | 2.27 | 2.79 | 3.76 |
| 1500 | 3 | 0.30 | 0.32 | 0.37 | 0.42 | 0.49 | 0.60 |
| | 4 | 0.55 | 0.60 | 0.68 | 0.77 | 0.90 | 1.12 |
| | 5 | 0.91 | 1.00 | 1.12 | 1.28 | 1.50 | 1.87 |
| | 6 | 1.30 | 1.42 | 1.60 | 1.84 | 2.16 | 2.69 |
| 2000 | 3 | 0.20 | 0.28 | 0.31 | 0.35 | 0.39 | 0.43 |
| | 4 | 0.48 | 0.52 | 0.58 | 0.64 | 0.73 | 0.87 |
| | 5 | 0.80 | 0.80 | 0.90 | 1.07 | 1.22 | 1.45 |
| | 6 | 1.14 | 1.23 | 1.37 | 1.54 | 1.76 | 2.10 |

For aspect ratio 2 : 1 area can be multiplied by 1.13
For aspect ration 3 : 1 area x 1.33

**Maximum permitted span for glass
supported on two edges (in m.)**

| wind load N/m$^2$ | glass thickness mm | angle of inclination from the horizontal | | | | | |
|---|---|---|---|---|---|---|---|
| | | 20° | 30° | 40° | 50° | 60° | 70° |
| 1000 | 3 | 0.33 | 0.35 | 0.38 | 0.41 | 0.45 | 0.52 |
| | 4 | 0.45 | 0.48 | 0.52 | 0.56 | 0.62 | 0.71 |
| | 5 | 0.58 | 0.6 | 0.66 | 0.72 | 0.79 | 0.92 |
| | 6 | 0.70 | 0.78 | 0.79 | 0.85 | 0.95 | 1.10 |
| 1500 | 3 | 0.31 | 0.32 | 0.34 | 0.37 | 0.40 | 0.44 |
| | 4 | 0.42 | 0.44 | 0.47 | 0.50 | 0.54 | 0.60 |
| | 5 | 0.54 | 0.57 | 0.60 | 0.64 | 0.70 | 0.78 |
| | 6 | 0.65 | 0.68 | 0.72 | 0.77 | 0.83 | 0.93 |
| 2000 | 3 | 0.29 | 0.30 | 0.32 | 0.33 | 0.36 | 0.39 |
| | 4 | 0.39 | 0.41 | 0.43 | 0.40 | 0.49 | 0.53 |
| | 5 | 0.51 | 0.53 | 0.50 | 0.59 | 0.63 | 0.68 |
| | 6 | 0.61 | 0.63 | 0.60 | 0.70 | 0.75 | 0.82 |

# APPENDIX 4 PIPE SIZING FOR THERMOSYPHONING SYSTEM

*Calculated maximum total lengths of straight galvanized steel piping and copper tubing for different heights of the storage tanks*

| No. of Absorbers | Assumed no. of fittings — tee | Assumed no. of fittings — 90° bend | Assumed no. of fittings — U- bend | Approx. height h (m) | Assumed method of connecting absorbers | Nominal pipe diameter (mm) | Equivalent length of pipe for fittings — 90° elbow (m) | Equivalent length of pipe for fittings — 90° bend (m) | Equivalent length of pipe for fittings — Gate valve (m) |
|---|---|---|---|---|---|---|---|---|---|
| 1 | | 4 | | 0,5 | 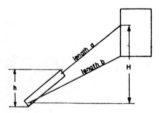 | 25 | 0,49 | 0,43 | 0,12 |
| | | | | | | 32 | 0,49 | 0,37 | 0,15 |
| 2 in parallel | 2 | 2 | | 0,5 | | 25 | 0,55 | 0,49 | 0,15 |
| | | | | | | 32 | 0,61 | 0,43 | 0,18 |
| | | | | | | 40 | 0,67 | 0,49 | 0,18 |
| 3 in parallel | 4 | 4 | | 1,5 | | 32 | 0,67 | 0,49 | 0,18 |
| | | | | | | 40 | 0,76 | 0,55 | 0,21 |
| | | | | | | 50 | 0,94 | 0,67 | 0,27 |
| 4 in parallel | 6 | 6 | | 0,5 | | 40 | 0,79 | 0,58 | 0,21 |
| | | | | | | 50 | 1,01 | 0,73 | 0,27 |
| 2 in series | | 4 | 1 | 1,0 | | 25 | 0,52 | 0,46 | 0,12 |
| | | | | | | 32 | 0,55 | 0,40 | 0,15 |
| 3 in series | | 4 | 2 | 1,5 | | 25 | 0,52 | 0,46 | 0,12 |
| | | | | | | 32 | 0,55 | 0,40 | 0,15 |
| 4 in series | | 4 | 3 | 2,0 | | 25 | 0,55 | 0,46 | 0,12 |
| | | | | | | 32 | 0,58 | 0,43 | 0,15 |
| 4 in series-parallel | 2 | 2 | 2 | 1,0 | | 32 | 0,64 | 0,46 | 0,18 |
| | | | | | | 40 | 0,73 | 0,52 | 0,21 |
| | | | | | | 50 | 0,88 | 0,64 | 0,24 |

## Note

(i) All results were calculated for a solar heater operating in Pretoria, with an absorber angle of inclination equal to 36°.

(ii) Total length of piping means length a plus length b.

(iii) The total number of fittings assumed for calculation purposes is given in columns 2, 3 and 4. Where extra fittings are inserted in the flow or return pipes, the maximum total length of piping must be reduced by the corresponding amount for each fitting, as given in columns 8, 9 and 10.

(iv) All calculations have been based on a 1,86m² absorber whose dimensions are 2,440m x 0,762m and whose long axis is horizontal.

(v) Values for copper tubing are given in brackets. Nominal diameters refer to outside diameters in this case.

(vi) Total lengths of piping in excess of about 15m are not recommended.

Max. total length of straight piping

| H=0,9m (m) | H=1,2m (m) | H=1,5m (m) | H=1,8m (m) | H=2,1m (m) | H=2,4m (m) | H=2,7m (m) | H=3,0m (m) |
|---|---|---|---|---|---|---|---|
| 4 (3) | 11 (8) | 17 (13) | 24 (19) | 31 (24) | 38 (30) | | |
| 17 (10) | 37 (23) | | | | | | |
| - | - | - | - | - | - | - | 6 (5) |
| 6 (2) | 16 (9) | 26 (15) | 36 (22) | | | | |
| 13 (8) | 32 (21) | | | | | | |
| - | - | - | 5 (-) | 7 (3) | 9 (5) | 12 (6) | 15 (8) |
| - | 5 (2) | 9 (6) | 15 (9) | 20 (13) | 25 (17) | 30 (21) | 35 (24) |
| 24 (20) | 56 (48) | | | | | | |
| - | - | - | 4 (-) | 7 (3) | 10 (6) | 13 (8) | 16 (10) |
| - | 8 (7) | 18 (16) | 27 (24) | 37 (33) | | | |
| - | - | 4 (2) | 12 (9) | 20 (15) | 27 (21) | 35 (27) | 43 (34) |
| - | - | 19 (11) | 42 (26) | | | | |
| - | - | - | - | 4 (2) | 12 (9) | 20 (16) | 29 (22) |
| - | - | - | - | 20 (11) | 45 (27) | | |
| - | - | - | - | - | - | 14 (10) | 24 (18) |
| - | - | - | - | - | 22 (13) | 51 (3) | |
| - | - | - | 4 (-) | 8 (3) | 13 (6) | 17 (9) | 22 (12) |
| - | - | 3 (-) | 13 (8) | 22 (15) | 31 (22) | 41 (29) | |
| - | - | 45 (38) | | | | | |

This table is reproduced from "Solar Water Heating in South Africa" by D.N.W. Chinnery. (C.S.I.R. Report No.248) Note that the values have been calculated for conditions pertaining in Pretoria and in less sunny climates somewhat shorter pipe runs and larger pipe diameters are advisable as there will be a smaller convective force to drive the circulation.

187

South (garden) elevation with solar panels installed.

*A.13 Any application for approval of alterations to a building is likely to require a locational drawing which shows clearly what is existing and what change is proposed.*

It is impossible to give any hard and fast rules about the requirements for either planning permission or Building Regulations approval with respect to solar water heating installations. This is because there are no national guidelines laid down by the Department of the Environment. Each local authority decides independently upon the laxity or stringency of its control over developments.

Many authorities I have had contact with have said that as the installation of solar water heaters generally involves no structural changes in the building, it does not require their consent. Others I have heard of, have been very demanding, and even obstructive. It is advisable therefore to contact your local planning department and building control section before starting work on your system, to find out just what the requirements are

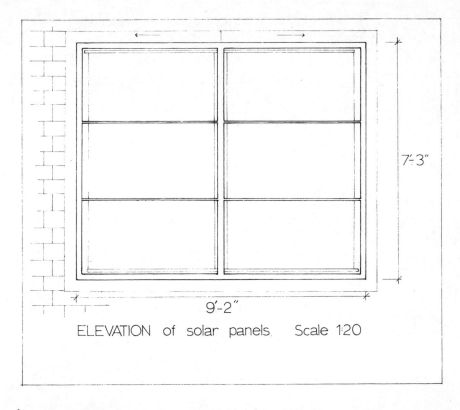

ELEVATION of solar panels. Scale 1:20

7'-3"

9'-2"

*A.14 The dimensions of any new constructions should be clearly marked.*

likely to be in your neighbourhood.

Unless you live in a listed historical building or in a conservation area, it is unlikely that you will need Planning Permission. If you do, you may have to prepare drawings or photographs indicating the changes your installation would effect upon the building's appearance and perhaps also a small map showing from where the changes will be visible.

For Building Regulations approval, the inspector will be more concerned with the stability of the collector mounting, its weight, the capacity of the storage tanks and the safety of the plumbing.

Officially, the local water authority should be notified of any changes in domestic plumbing systems. They will want to know the plumbing layout, pipe and tank sizes, and the materials used.

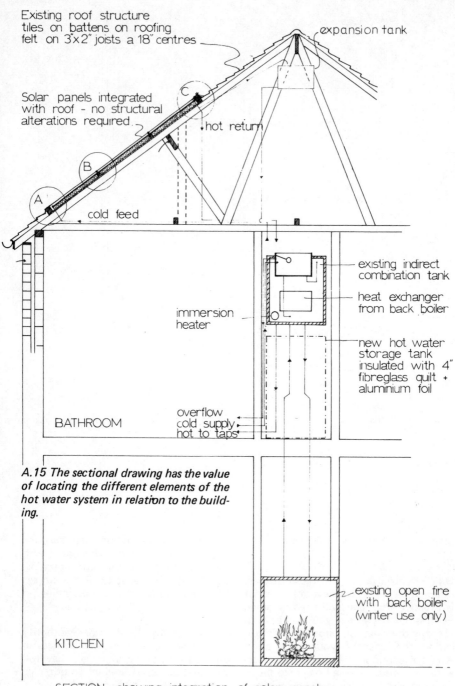

Existing roof structure
tiles on battens on roofing
felt on 3"x2" joists a 18" centres

expansion tank

Solar panels integrated
with roof - no structural
alterations required

C

hot return

B

A

cold feed

existing indirect
combination tank

heat exchanger
from back boiler

immersion
heater

new hot water
storage tank
insulated with 4"
fibreglass quilt +
aluminium foil

overflow
cold supply
hot to taps

BATHROOM

**A.15 The sectional drawing has the value
of locating the different elements of the
hot water system in relation to the building.**

existing open fire
with back boiler
(winter use only)

KITCHEN

SECTION showing integration of solar panels
with existing plumbing system. Scale 1:20

# NOTES.

1. The integration of solar panels involves no structural alterations to the roof. Tiles only are removed.

2. The new solar hot water tank is placed under the existing indirect combination tank, so there are no internal alterations required other than achieving plumbing connections.

3. The 6 sq m. of solar panel will provide all domestic hot water requirements for 7-9 months of the year.

4. The increased hot water storage capacity and the much improved insulation of the tanks will cover a dull period of 4-6 days of negligable radiation gains.

5. The solar heating circuit can be isolated from the rest of the plumbing system.

6. Solar panels are sited on the side of the roof above the bathroom, kitchen and water tanks to achieve short pipe runs.

7. Being on the back garden face of the roof, the panels will be unobtrusive (and in fact only visible from the bottom of the garden.)

*A.16 Clearly printed or typed notes accompanying the drawings can help to clarify them and serve to emphasise how little the solar system will alter the existing amenity and appearance.*

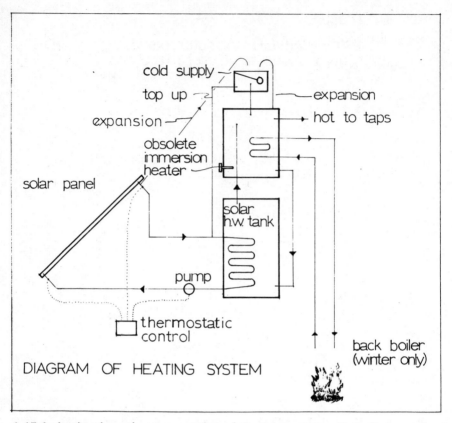

cold supply

top up

expansion

expansion

hot to taps

obsolete
immersion
heater

solar panel

solar
h.w. tank

pump

thermostatic
control

back boiler
(winter only)

DIAGRAM OF HEATING SYSTEM

*A.17 A simple schematic representation of the solar system will clarify its mode of operation to authorities who may not be familiar with its working. (Refer to ch.7 and ch.9 when designing your own layout.)*

The series of drawings in this appendix were prepared for a family who wished to install a solar system in their rented council house. They did not require planning permission but they did need the approval of the estate managers. The information presented by the drawings is comparable to that which would be needed in an application for planning permission or Building Regulation approval, and in this case it was sufficient to obtain consent.

If you are refused permission, ask why. If it is on grounds of safety, or failure to comply with a particular regulation, seek the building inspector's advice on what changes are necessary in your design. If refusal is based upon objections to the appearance of the collectors, enquire how this might be improved and consider alternative, and less obtrusive mounting locations. If the problem persists, contact local amenity groups like the Conservation Society. They are

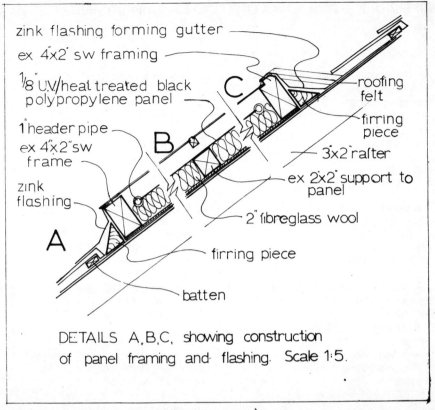

zink flashing forming gutter

ex 4"x2" sw framing

⅛" UV/heat treated black polypropylene panel

C

roofing felt

1" header pipe

ex 4"x2" sw frame

B

firring piece

3x2" rafter

zink flashing

ex 2"x2" support to panel

2" fibreglass wool

A

firring piece

batten

DETAILS A,B,C, showing construction of panel framing and flashing. Scale 1:5.

*A.18 A section drawn through the mounted collector will help the inspector to assess its stability and the danger of rainwater leakages through the roof.*

more likely to see the links between solar systems, reduced fossil fuel usage and consequent environental improvement. If you can convince them that the solar installation represents an attack on one of the roots of environmental despoilation, and therefore merits exception from the authority's arbitrary aesthetic rulings, you will gain a valuable ally in presenting your case.

Finally, if you run up against a brick wall of bureaucracy, contact your local newspaper. Solar energy is news and any local authority which is unreasonably obstructing individuals in their attempts to reduce their energy consumption, deserve all the bad publicity that comes their way.

# APPENDIX 6   LIST OF SOLAR COLLECTOR MANUFACTURERS

**Air Distribution Equipment (M&W) Ltd**
64 Whitebarn Road
Llanishen, Cardiff, CF4 5HB
Tel. Cardiff (0222) 40404

**Alcoa Ltd**
Waunarlwydd Works
Gowerton, Swansea
Tel. Gowerton 783301

**Allseasons**
Solar Heating Centre
20 Baldock Street
Ware, Herts
Tel. Ware 61635

**Aluglaze (Alcan Booth Systems)**
Raans Road
Amersham, Bucks
Tel. Amersham (02403) 21262

**Ansco Solar Panels Ltd**
61 Cornwall Street
Birmingham
Tel. (021) 236 5677/6963

**Antarim Ltd**
Solar Works
New Street, Petworth, Sussex
Tel. (0798) 43005

**Asahi Trading Co Ltd**
Asahi House, Church Road
Port Erin, Isle of Man
Tel. Port Erin (062483) 3379 & 3758

**Avica Equipment Ltd**
Mark Road, Hemel Hempstead
Tel. Hemel Hempstead 64711
(In association with Engineering &
Marine Products)

**Bexley Glass Ltd**
37 High Street
Bexley, Kent, DA5 1AB
Tel. Crayford 53311—4

**Bifurcated Engineering Ltd**
P.O. Box 2, Mandeville Road
Aylesbury, Bucks, HP21 8AB
Tel. Aylesbury 5911

**Building Heat Conservation**
191 Vale Road, Tonbridge
Tel. Tonbridge 352120

**Calor Sol Ltd**
Lancaster Road
Harlescott Grange,
Shrewsbury SY1 3NG
Tel. Shrewsbury 51578/9
(Sub. of A.T. Marston & Co Ltd)

**Cheesman Solar Systems**
6 High Street, Stotfold, Hitchin, Herts
Tel. Hitchin 730134

**Commercial Solar Energy**
16a Pelham Road
Sherwood, Nottingham
Tel. (0602) 601847

**Consumer Power Co Ltd**
Market House, 1 Market Street
Saffron Walden, Essex CB10 1JB
Tel. (0799) 22220

**Design & Materials Ltd**
Carlton in Lindrick Industrial Estate
Worksop, Notts S81 91B
Tel. (0909) 730333

**District & County Installations**
Solar Development Division
74 Rutland Street
Leicester LE1 1SB
Tel. (0533) 28925

**Distrimex Ltd**
88 The Avenue
London, NW6 7NN
Tel. 01 459-1391

**Don Engineering (South West) Ltd**
Wellington Trading Estate
Wellington, Somerset TA21 8SS
Tel. (082347) 3181

**Drake & Fletcher Ltd**
Swimming Pools
Parkwood, Sulton Road
Parkwood, Kent, ME15 9NW
Tel. Maidstone 55531

**Electra Air Cond. Services Ltd**
66-68 George Street
London W1 5RG
Tel. 01 487 5606

**Engineering & Marine Products Ltd**
Solar Energy Division
Windmill Grove, Wicor Mill Lane
Dorchester, Dorset
Tel. Cosham 74873

**4—T Engineering Ltd**
North Dock
Llanelli, Dyfed
Tel. (05542) 75041/2

**Grumman International Inc.**
64-65 Grosvenor Street
London W1X DB
Tel. 01 629-3847

**I.M.I. Range Ltd**
P.O. Box 1
Stalybridge, Cheshire SK15 1PQ
Tel. (061) 338 3353

**F. J. Jones**
Swimming Pools
30 Gorsty Hill Road, Rowley Regis
Warley, W. Midlands
Tel. (021) 559 3658

**Kenelek**
2 Wellington Road
Bournemouth, Dorset BH8 8JN
Tel. (0202) 20704/22100

**Kent Solar Traps Ltd**
10 Albion Way  Maidstone, Kent
Tel. Maidstone 674041

**Kleen Line Systems**
School Lane, Knowsley
Prescot, Merseyside
Tel. (051) 546-8225

**Lennox Industries Ltd**
P.O. Box 43, Lister Road
Basingstoke, Hants RG22 4AR
Tel. Basingstoke 61261

**McKee Solaronics Ltd**
12 Queenborough Road
Southminster, Essex
Tel. Maldon 772477

**Natural Power (UK) Ltd**
Yorkshire House
Greek Street, Leeds LS1 5SX
Tel. (0532) 468146/7

**D. J. Neil Ltd**
P.O. Box 31
Macclesfield
Chesire SK10 2EX
Tel. 0625 20179

**N.H.C. Solar Heat**
Longridge Trading Estate, Longridge
Knutsford, Cheshire, WA16 7BR
Tel. Knutsford 53411

**Oceanware Ltd**
Herons Reach, Cound Moor
Nr Shrewsbury, Salop
Tel. Acton Burnel 648

**Paltec International Ltd**
3 Rue des Mielles
St. Helier, Jersey, C.I.
Tel. (0534) 71224

**Production Methods Ltd**
West Arthurlie Works
Barrhead
Glasgow G78 1LQ
Tel. (041) 881 2281

**Redpoint Associates Ltd**
Cheyney Manor
Swindon, Wilts, SN2 2QN
Tel. (0793) 28440

**Refrigeration Appliances Ltd**
Haverhill, Suffolk, CB9 8PT
Tel. (0440) 2653/7

**Robinsons Developments Ltd**
Swimming Pools
Robinson House
Winnall Industrial Estate
Winchester, SO23 8LH
Tel. (0962) 61777

**Senior Platecoil Ltd**
Otterspool Way, Watford Bypass
Watford, Herts
Tel. Watford 26091/35571

**Solar Apparatus & Equipment Ltd**
Brunel Road
Newton Abbot, Devon
Tel. Newton Abbot 3003

**Solar Centre**
176 Ifield Road
Chelsea, London SW10
Tel. 01 370 4804

**Solar Economy**
Balksbury Hill, Upper Clatford
Nr Andover, Hants
Tel. (0264) 51522

**Solar Energy Industries**
4 Fleming Close
London, SW10 0AH

**Solar Energy Products**
22 James Hall Gardens
Deal, Kent
Tel. Deal 63486

**Solar Energy Services**
134 High Street
Deritend, Birmingham, B12 0JU

**Solargen Solar Heating**
Hadfield Industrial Estate
Waterside, Hadfield
Hyde, Cheshire
Tel. Glossop 63131

**Solar Heat Ltd**
99 Middleton Hall Road
Kings Norton, Birmingham 30
Tel. (021) 458 1327

**Solarplan**
St Albans House
577-579 Harehills Lane, Leeds 9
(Tel. not known)

**Solar Power Systems (Bradford) Ltd**
935 Leeds Road, Bradford
Tel. (0274) 662536

**Solartherm Energy Systems**
396 Cheetham Hill Road
Manchester 8
Tel. (061) 795 0717

**Solar Water Heaters Ltd**
Swimming Pools
153 Sunbridge Road
Bradford, Yorks, BD1 2PA
Tel. (0274) 24664

**Solchauf**
Unit 5
Rashes Green Industrial Estate
Dereham, Norfolk
Tel. Dereham 2114

**Stellar Heat Systems Ltd**
Stellar House
Bath Road, Keynsham SB18 1TN
Tel. Keynsham 67111

**Sunheat Systems Ltd**
Bayshill House, Bayshill Road
Cheltenham, Glos.
Tel. (0242) 34852

**Sunmaster (Maybeck) Ltd**
Foster House
Studley, Warwicks
Tel. Studley 2454 & 3833

**Sunsense Ltd**
1 Church Street
Northborough, Peterborough
Tel. Peterborough 252 or 672

**Sunwarm Solar Systems
(Marketing) Ltd**
189 High Street, Boston Spa
Wetherby, Yorks
Tel. Boston Spa 844800

**Surrey Solar Heating (Bookham) Ltd**
241 Lower Road
Bookham, Surrey
(Tel. not known)

**Tetramarl (Solar Energy) Ltd**
32A High Road
South Benfleet
Essex SS7 5LH
Tel. 03745 3381

**Thermal Energy Components Ltd**
Mushroom Farm Industrial Estate
Derby Road
Eastwood, Nottingham
Tel. Langley Mill 69579

**Thermal Services**
17 Queen Street, Stotfold
Hitchin, Herts
Tel. Hitchin 731974

**U.F. Foam Consultants**
184 Cedar Road
Earl Shilton, Leics.
Tel. (0455) 42965

**Verdik Ltd**
Colliers Corner, Lane End
Nr High Wycombe, Bucks
Tel. High Wycombe 881254

**Washington Engineering Ltd**
Industrial Road
Washington, Tyne & Wear
Tel. (0632) 463001

**Yorkshire Solar Systems**
387 York Road, Leeds
Tel. Leeds 486392

# INDEX OF MANUFACTURERS OF USEFUL MATERIALS AND EQUIPMENT

*Airwrap* (A matrix of air bubbles in a polythene skin)
Abbots Packaging Ltd.
Gordon House,
Oakleigh Road South,
London N11.
Tel: 01 368 1266

*Anti-freeze Solution* see Fernox FPI

*Amkit* (Nylon pipe suppliers)
Autocon Manufacturing Co.,
Spring House,
10 Spring Place,
London NW5 3BH.
Tel: 01 485 9328

*Butyl*
Butyl Products Ltd.,
11 Radford Crescent,
Billericay,
Essex,
Tel: Billericay 53281

*Controls* (Electronic differential temperature controllers for pumps)
  (i)  Air Distribution Equipment Ltd.,
      64 Whitebarn Road,
      Llanishen, Cardiff CF4 5HB
      Tel: Cardiff 0222 40404
  (ii)  Don Engineering (S.W.) Ltd.,
      Wellington Industrial Estate,
      Wellington, Somerset TA21 8SS
      Tel: Wellington (0823 47) 3181
  (iii)  Sunsense Ltd.,
      1 Church Street,
      Northborough,
      Peterborough
      Tel: Peterborough 252 or 672.

*Copper Cylinders* see IMI

*Corex* (polypropylene wafer packaging material)
Corruplast Ltd.,
1 New Oxford St.,
London WC1.
Tel: 01 405 8797

*Dow Corning Silicone Sealant*
Hanley Bros.,
Heathrow House,
Bath Rd.,
Hounslow.
Tel: 01 941 1100 or 01 759 2600

*Diverter Valve* See Honeywell

*Fibre Glass Tanks*
BTR Reinforced plastics,
Barnsfield Place,
Rockingham Rd.,
Uxbridge,
Middlesex UB8 2UL.
Tel: Uxbridge 933360

*Fibre Glass Cylinders*
Enviromark Ltd.,
Kraemar Division,
Evingar Rd.,
Whitchurch,
Hampshire.
Tel: Whitchurch (025 682) 2162

*Fernox FP-1* (anti-freeze)
Industrial (Anti-Corrosion) Services,
Britannica House,
214-224 High Street,
Waltham Cross,
Herts.
Tel: Waltham Cross 28355

*Fortec* (combination hot water cylinder) see IMI

*Harcopak* (Combination hot and cold tanks)
Harvey Fabrication Ltd.,
Woolwich Rd.,
London SE7 7RJ
Tel: 01 858 3232.

*Heat Exchangers*
Hot Rod (finned tube)
Kleenline Systems,
School Lane,
Knowsley.
Prescot,
Merseyside
Tel: 051 546 8225
See also Micraversion and IMI

*Heat Pipes*
(i)  Redpoint Associates,
     Cheyney Manor,
     Swindon,
     Wilts. SN2 2QN
     Tel: 0793 28440
(ii) Isoterix Ltd.,
     1 Bank House,
     Balcombe Rd.,
     Horley, Surrey.
     Tel: Horley 2099

*Honeywell* (3-way divertor-valve with thermostatic control)
Honeywell Ltd.,
Honeywell House,
Charles Square,
Bracknell, Berks.
Tel: Bracknell 24555

*IMI Range Ltd* (Manufacturers of copper cylinders, heat exchangers and solar collectors)
Stalybridge,
Cheshire, SK15 1PQ
Tel: 061 338 3353

*Micraversion* (Immersion-type heat exchanger)
Wednesbury Tube Co.,
Oxford St.,
Bilston,
West Midland V14 7DS
Tel: 0902 41133

*Nylon Pipe* (see also Amkit)
Solar Heat Panels,
8 St. Georges Walk,
Allhallows,
Medway,
Kent ME3 9P
Tel: Medway (06347 271 595)

*Polypropylene Balls* (floating insulation in open-top tanks)
Allplas Balls from
Capricorn Industrial Services Ltd.,
49 St. James Street,
London SW1A 1JY
Tel: 01 493 8847

*Pressurized Vessels & Filling Assemblies*
Kleenline Systems,

School Lane,
Knowsley,
Prescot,
Merseyside
Tel: 051 546 8225

*Sidewinder* (heat exchanger) see IMI

*Silicone Sealant* see also Dow Corning.
Silicone Products,
Silaicon House,
Cranfield Road,
Lostock,
Bolton BL6 4QD
Tel: Horwich 692531

*Solar cells* (Photo-voltaic devices for the generation of electricity)
(i) Ferranti Ltd.,
    Electronic Components Division
    Gem Mill, Chadderton,
    Oldham, Lancs.
    Tel: 061 624 0515
(ii) Joseph Lucas,
    Gt. Hampton Street,
    Birmingham BI8 6AH
    Tel: 021 236 5050
(iii) Solar Power Ltd.,
    Burton,
    Norwich NR6 6AX
    Tel: Norwich 44914

*Tedlar* (u.v. resistant plastic film with high solar transmission)
Russell-Cowan Properties Ltd.,
Solar Energy Division,
70 Courtfield Gardens,
London SW5
Tel: 01 370 4804

*Thermometers*
Brannan Thermometers Ltd.,
Cleator Moor,
Cumbria,
Tel: 0946 810413

*Thermal Paste*
Industrial Science Ltd.,
Leader House,
Shargate Road,
Dover.
Tel: 0304 202656

**Direct Use of the Sun's Energy**
Farrington Daniels
Ballantine Paperback, 1964. £1.00

Perhaps the most popular book written on solar energy, it presents an over-view of all lines of research aimed at harnessing the sun's energy. It is richly annotated with references for those who wish to follow up particular interests in more detail, but written in simple language which can be easily assimilated by non-specialists.

**Sun Power**
J. C. McVeigh
Pergamon, 1977. £2.75

A more up to date over-view than Farrington Daniels, this is an excellent starting point for the student who wishes to delve into the many applications of solar energy. It draws successfully on the author's personal experience with low temperature applications and his familiarity with international research.

**Solar Energy & Building**
S. V. Szokolay
Architectural Press, 1975. £6.95

Especially useful for designers, this work incorporates thirty-nine case studies of solar houses from all over the world and closes with a step-by-step design guide. There is by the way some controversy over the very large deviations from south which Szokolay suggests are possible without greatly reducing the returns from a solar system. Such quibbles aside, I have found this a valuable reference book.

**Sun Spots**
Steve Baer
Available from Compendium Books, Camden High Street, London. £2.10

Baer is well known for his innovative solar devices and his book is equally innovative. Sections on heat exchangers and rock pile storage are interlaid with conspiracy theories and visions of state intervention to interfere with indi-viduals' access to sunlight. Sometimes fictional sequences serve as vivid illustrations of the factual information. The story of the petrol-drinking addict for example, is a strong reminder of the importance of energy in our daily lives, and its role as a unifying principle.

**Solar Energy for Man**
B. J. Brinkworth
The Compton Press, 1972. £3.50

If you become really intrigued by solar energy, it is well worth reading this book. It is more a students text than the other books mentioned, but care has been taken to jog the memories of those who must struggle to remember their school physics. The effort is amply rewarded by the satisfaction of estab-lishing a clear theoretical understanding of the various physical processes which take place between nuclear explosions in the heart sun and your bathful of hot water.

**The Solar Home Book**
Bruce Anderson
Prism Press, 1976, £5.00.

This is probably the most readable

account of how to use solar energy for heating the home. It places a welcome emphasis on simple passive techniques whilst still detailing the more expensive mechanical solutions using collector/ storage circulation loops. Written primarily for those who design or build houses, the layman is catered for by features on DIY projects and by graphical explanations of the technical terms and concepts which are mentioned.

### The Autonomous House
Brenda & Robert Vale
Thames & Hudson, 1975. £2.50

I am always grateful to the Vales for being my first source of information when I became interested in the idea of using alternative energy sources. Their reports were among the first items of technical literature to appear following the upsurge of interest in self-sufficiency in the early seventies, and this book, remarkable in its scope, has arisen from their pioneering work.

### Technological Self-Sufficiency
Robin Clarke
Faber & Faber, 1976. £2.95

Not so much a how-to-do-it book as a yes-you-can-do-it book. Robin Clarke's experience in changing from city editor to rural jack-of-all-trades make interesting reading and reassure that you do not need to be a gifted craftsman to take control of your physical living environment.

### Solar Greenhouses
R. Fisher & Bill Yanda
John Muir Publications, Santa Fe, New Mexico. 1976. $6.00
(available from Compendium Books, London)

This is the other way the DIY enthusiast can harness the sun. A greenhouse on the south side of your house can contribute useful calories to your space heating as well as your diet. Not so much

attention has been paid to this attractive method of solar energy utilisation since the Victorian vogue for botanical glass-houses and lean-to conservatories. Perhaps this book will help remind us of our valuable heritage.

### Soft Energy Paths
Amory Lovins
Penguin Books, 1977. £0.95

A welcome reassurance to society dithering on the brink of change. By means of a rational appraisal of all our potential energy resources, from the "hard" solutions like nuclear power, to the "soft technologies" utilising the sun, wind and biomass conversion, Lovins demonstrates that the soft paths are not only safer, but also cheaper. Those who have kept themselves outside the arguments on energy policy because of the tendency for debates to become lost in technical intricacies, will be pleased to read Lovins' view that: "The basic issues in energy strategy, far from being too complex and technical for ordinary people to understand, are on the contrary too simple and political for the experts to understand."

### The Energy Question
Gerald Foley
Pelican Books, 1976. £0.90

As a former student of the author, I have been considerably influenced by the ideas expressed in this book. Gerry Foley is one of those people who can casually predict Armageddon then giggle, leaving the impression that he is completely detached, or perhaps has the solution up his sleeve. As the title suggests however, it is not a book of answers more a guide for those who are looking for the questions. It will also help the reader to evaluate more critically many of the glib "answers" which have been put forward in the past.

**Keeping Warm for Half the Cost**
Colesby and Townsend,
Prism Press, 1976 £1.50.

Insulation is the most urgent and important first step in reducing our dependance upon non-renewable energy sources. It also saves money. This book is a clear guide to household insulation and if you do not already have such a book, buy this before going any further in a search for alternative energy sources.

**The Self Help House Repair Manual**
Andrew Ingham
Penguin, 1975. £0.60

A book I refer to continually. An essential handbook for anyone who has to tackle roofing, gas installation, plumbing wiring, i.e. practically any household building work. It is outstandingly good value, fully illustrated and organised in carefully indexed and cross-referenced sections.

**The Readers Digest Repair Manual —**
**A Complete Guide to Home Maintenance**
The Readers Digest Co.

A much plusher guide to household repairs. It has a much wider scope than Ingham's book, including bicycle repair and mending broken washing machines, but it stops short of giving details for bigger jobs and advises calling in tradesmen.

**Zen & the Art of Motorcycle**
**Maintenance**
Robert Pirsig
Corgi, 1976 (1974). £0.95

This has reintroduced me to the value of the Western way of reasoning. The reaction to the evils of modern technology has sometimes threatened to throw out the baby with the bath water. Apart from all the other things which might be said about it, this book offers the philosophical basis which will allow you to remain calm and methodically seek out the solution when the solar system you just broke your back installing refuses to circulate for no apparent reason.

## APPENDIX 8 ORGANISATIONS AND ADDRESSES

### UK—ISES

The UK section of the International Solar Energy Society has the aim of promoting interest and understanding of the use of solar energy. No special qualifications are required for membership which costs £12.50/year (£7.50 for students). Members receive both the international and national magazines and generous discounts on the society's many publications and conference fees. Affiliate members pay a reduced subscription of £7.50 but do not receive the international magazine.
Publications include:
Solar Energy — A UK Assessment. 1976
£10.00 (£5.00 to members)
Flat Plate Collectors & Solar Water Heating
£2.00 (£1.00 to members)
UK—ISES, c/o The Royal Institution, 21 Albemarle Street, London W1X 4BS
Tel. 01 493 6601

### NCAT

The National Centre for Alternative Technology is a large demonstration plot and permanent exhibition of windmills, solar collectors, heavily insulated housing, methane gas production and water power. They produce various information sheets and the centre is open to the public on payment of an admission charge.
NCAT, Machynlleth, Powys, Wales.

### The Copper Development Association

They can provide advice on the use and availability of copper for solar systems.
CDA, Orchard House, Mutton Lane, Potters Bar, Herts., EN6 3AP
Tel. Potters Bar 50711

### The Aluminium Federation

Broadway House, Calthorpe Road, Birmingham B15 1TN.
Tel. 021 455 0311

### Pilkington — Solar Energy Advisory Service

Its aim is to promote the development and application of practical devices and architectural designs for the use of solar energy in buildings and industry. It offers information and advice on all aspects of solar utilization and seeks to match the products of the glass industry to the rapidly increasing exploitation of solar energy.
Solar Energy Advisory Service, Pilkington Bros. Ltd., Prescot Road, St. Helens, Merseyside WA10 3TT.
Tel. St. Helens (0744) 28882 x2272

### RTG

The Rational Technology Group is a small co-operative to which the author belongs. It offers design, research and educational services and is particularly enthusiastic about working with and for the developments of housing and industrial co-operatives, seeing the control of technology being as important as the rational choice of appropriate hardware.
45—47 Brunel Road, Rotherhithe, London SE16. Tel: 01-263-3246.

### Country College

Country is a privately run organization with its HQ in a Hertfordshire cottage. It exists to encourage sustainable lifestyles and to this effect holds lectures, seminars, produces films, publications and has organised extremely successful exhibitions on both solar collectors and wood-burning stoves.
Country College, 11 Harmer Green Lane, Digswell, Welwyn, Herts AL6 0AY
Tel. Welwyn (043871) 6367

# APPENDIX 9   UNITS AND CONVERSION FACTORS

## UNITS

### Length

Throughout the text, measurements of length have been given in metric units with imperial units in brackets.

1 micrometre   ($\mu$m) = 1000 nanometres (nm)
1 millimetre (mm) = 1000 $\mu$m
1 centimetre (cm) = 10 mm
1 metre (m) = 100 cm

| Multiply | By | To Obtain |
|---|---|---|
| centimetres (cm) | 0.3937 | inches |
| inches | 2.54 | centimetres (cm) |
| metres (m) | 3.28 | feet (ft.) |
| feet | 0.3048 | metres |

### Plumbing Sizes

With the exception of threaded connections, plumbing components are now marketed in metric sizes. In some cases these are slightly different from the imperial sizes they replace and adapters are sometimes required when joining old pipework to modern pipes. The nominal equivalents for pipe sizes are as follows:

½ inch − 15 mm          1 inch − 28 mm
¾ inch − 22 mm          1¼ inch − 35 mm
                        1½ inch − 42 mm

### Area

| Multiply | By | To Obtain |
|---|---|---|
| sq. metres | 10.764 | sq. feet |
| sq. feet | 0.0929 | sq. metres |

206

## Volume

Liquid quantities have been quoted in litres with imperial gallons (UK) in brackets.

1 millilitres (ml.) = 1 cubic centimetre (cc)
1 litres = 1000 ml.
1 cubic metre = 1000 litres

| Multiply | By | To Obtain |
|---|---|---|
| litres | 0.22 | gallons (UK) |
| litres | 0.2642 | gallons (US) |
| gallons (UK) | 4.546 | litres |
| gallons (US) | 3.7853 | litres |
| cubic metres | 35.3147 | cubic feet |
| cubic feet | 0.0283 | cubic metres |

## Weight & Water

Water expands when it is heated and also as it approaches freezing. The relationship between volume and weight therefore is different for different temperatures. The following equivalents are easily remembered round numbers which provide a close guide.

1 cc. of water weighs 1 gram.
1 litre of water weighs 1 kilogram (kg.)
1 gallon (UK) of water weighs approximately 10 pounds.

## Temperature

Degrees Celsius (deg.C.), which are the same as degrees Centigrade, have been used throughout the book with degrees Fahrenheit (deg.F.) shown in brackets.

to convert deg.C. to deg. F. : $deg.C. = \dfrac{5(deg.F-32)}{9}$

to convert deg.F. to deg. C.: $deg.F. = \dfrac{9 \times deg.C.}{5} + 32$

1 scale degree Celsius = 1.8 scale degrees Fahrenheit

sample equivalents:

| deg.F. | deg.C. | |
|---|---|---|
| | -273.15 | Absolute zero ($0^o$ Kelvin) |
| 32 | 0 | freezing point of water |
| 50 | 10 | |

| | | |
|---|---|---|
| 59 | 15 | cold tap water |
| 64 | 18 | |
| 70 | 21 | comfortable room air temperature |
| 98.2 | 36.8 | body temperature |
| 107 | 42 | |
| 113 | 45 | hot bath temperature |
| 131 | 55 | hot tap temperature |
| 212 | 100 | boiling point of water |

## Work

Work is a familiar word but for scientific purposes we need a precise definition such as — work is that which is done when an object is moved or altered. Work can be measured in terms of the energy used to accomplish the task of altering an object. See energy units.

## Power

Power is the rate at which work is done which is the same as saying that power is the rate at which energy flows. The most internationally recognised unit of power is the "watt" (w.). Formerly "horsepower" (h.p.) was a more familiar term. The watt is a very small unit and the kilowatt (kW.) is often used instead.

1 kW. = 1000 w.

| Multiply | By | To Obtain |
|---|---|---|
| kW. | 1.34 | hp. |
| hp. | 0.7463 | kW. |

## Energy

Because energy is apparent in all the activity we see around us, jumping, cooking, eating, breathing and watching TV to name a few, a great diversity of units have been invented to measure it, e.g. foot-pounds, therms, calories and kilowatt-hours. Such diversity can of course lead to confusion and there have been attempts to standardise with a single unit. The System International (SI units) use the joule which is a power of one watt applied for one second. In this book however, the kilowatt-hour (kWh.) is used throughout. This has been chosen simply because of its familiarity and the ease with which it can be conceptualized. Just as the horse-power was an easily grasped concept two centuries ago, today we are all acquainted with the kWh., the unit of electricity on which our power bills are based. A kWh. can be thought of as the amount of energy, in the form of heat, which is released when a normal one bar electric fire is allowed to burn for one hour.

All energy units can be interchanged although it may sometimes seem amusing to see units used in unaccustomed contexts. The dietician for example chooses to describe the energy content of food in terms of calories. It would however be quite accurate to say that the well-fed western man eats 3 kWh. per day. Similarly we might define his work output as being a maximum of 0.5 kWh/day. This of course makes obvious our rather low machine efficiency. Could it be embarrassment that makes us continue to hide our basic energy exchanges behind a baffling variety of units?

1 kWh. = 1000 watt-hours (wh.)

| Multiply | By | To Obtain |
|---|---|---|
| kWh. | 3600 | kilojoules |
| kWh. | 859.845 | kilocalories (kcal) |
| kWh. | 3412 | British thermal units (Btu.) |
| kWh. | 1.34 | horsepower-hours |
| kWh. | 0.034 | therms |
| kWh. | 1980 000 | foot-pounds |
| kilojoules (kJ) | 0.278 | watt-hours |
| kilocalories (kcal.) | 1.16 | watt-hours |
| British thermal units (Btu.) | 0.293 | watt-hours |
| horsepower-hours | 0.746 | kWh. |
| therms | 29.4 | kWh. |

## Energy & Temperature

Approximately 5kWh. are required to raise the temperature of

100 litres of cold water (at 12 deg.C) to hot tap temperature of 55 deg.C.
1.16 wh. raises the temperature of 1 litre of water by 1 deg.C.
1 calorie raises the temperature of 1 cc. of water by 1 deg.C.
1 Btu. raises the temperature of 1 pound of water by 1 deg. F.

## Rates of flow

1 kg/min. = 1 litre/min.

| Multiply | By | To Obtain |
|---|---|---|
| litres/min. | 13.20 | galls (UK)/hour |
| galls (UK)/hr | 0.076 | litres/min. |
| galls(UK)/hr | 4.55 | litres/hr. |

## Questionnaire

This book has been written in the belief that, given sufficient information, people can literally tap the sun's energy. In order to assess the potential for DIY solar heating, I have compiled a short questionnaire. I would be very grateful to readers who take the trouble to complete this and send it to:

Practical Solar Heating, Prism Press, Stable Court, Chalmington, Nr. Dorchester, Dorset. DT2 0HB.

**Yes    No**

....... ....... I have a solar water heating system.

....... ....... I have installed the plumbing myself.

....... ....... I have mounted the collectors myself.

....... ....... I have constructed the collectors myself.

....... ....... I have used radiator panels in the collector.

....... ....... I have used copper tube and sheet in the collector.

....... ....... I have used a different type of collector.

The collector area is       sq.metres (    sq.ft.)
The solar tank volume is     litres      (    gallons).

**Yes    No**

....... ....... I intend to have a commercial solar system installed.

....... ....... I intend to buy a commercial system and install it myself.

....... ....... I intend to construct collectors and install them myself.

**Yes    No**

           I do not intend to install a solar water heating system. Reasons:

....... ....... a) They are too expensive.

....... ....... b) I need more information.

....... ....... c) It requires too much work.

....... ....... d) My home is unsuitable for fitting the equipment.

Comments:


(Optional)
Name . . . . . . . . . . . . . . . . . . . . . . . . . . . . . . . . . . . .
Address . . . . . . . . . . . . . . . . . . . . . . . . . . . . . . . . . . .
. . . . . . . . . . . . . . . . . . . . . . . . . . . . . . . . . . . . . . . .

## Questionnaire

This book has been written in the belief that, given sufficient information, people can literally tap the sun's energy. In order to assess the potential for DIY solar heating, I have compiled a short questionnaire. I would be very grateful to readers who take the trouble to complete this and send it to:

Practical Solar Heating, Prism Press, Stable Court, Chalmington, Nr. Dorchester, Dorset. DT2 0HB.

| Yes | No | |
|-----|-----|-----|
| ....... | ....... | I have a solar water heating system. |
| ....... | ....... | I have installed the plumbing myself. |
| ....... | ....... | I have mounted the collectors myself. |
| ....... | ....... | I have constructed the collectors myself. |
| ....... | ....... | I have used radiator panels in the collector. |
| ....... | ....... | I have used copper tube and sheet in the collector. |
| ....... | ....... | I have used a different type of collector. |

The collector area is          sq.metres    (      sq.ft.)
The solar tank volume is       litres       (      gallons)

| Yes | No | |
|-----|-----|-----|
| ....... | ....... | I intend to have a commercial solar system installed. |
| ....... | ....... | I intend to buy a commercial system and install it myself. |
| ....... | ....... | I intend to construct collectors and install them myself. |

**Yes    No**

I do not intend to install a solar water heating system. Reasons:

| Yes | No | |
|-----|-----|-----|
| ....... | ....... | a) They are too expensive. |
| ....... | ....... | b) I need more information. |
| ....... | ....... | c) It requires too much work. |
| ....... | ....... | d) My home is unsuitable for fitting the equipment. |

Comments:



(Optional)

Name . . . . . . . . . . . . . . . . . . . . . . . . . . . . . . . . .

Address . . . . . . . . . . . . . . . . . . . . . . . . . . . . . . .

. . . . . . . . . . . . . . . . . . . . . . . . . . . . . . . . . .